CATCHING STARDUST

Also available in the Bloomsbury Sigma series:

Sex on Earth by Jules Howard
p53: The Gene that Cracked the Cancer Code by Sue Armstrong
Atoms Under the Floorboards by Chris Woodford
Spirals in Time by Helen Scales
Chilled by Tom Jackson
A is for Arsenic by Kathryn Harkup
Breaking the Chains of Gravity by Amy Shira Teitel
Suspicious Minds by Rob Brotherton
Herding Hemingway's Cats by Kat Arney
Electronic Dreams by Tom Lean
Sorting the Beef from the Bull by Richard Evershed and Nicola Temple
Death on Earth by Jules Howard
The Tyrannosaur Chronicles by David Hone
Soccermatics by David Sumpter
Big Data by Timandra Harkness
Goldilocks and the Water Bears by Louisa Preston
Science and the City by Laurie Winkless
Bring Back the King by Helen Pilcher
Furry Logic by Matin Durrani and Liz Kalaugher
Built on Bones by Brenna Hassett
My European Family by Karin Bojs
4th Rock from the Sun by Nicky Jenner
Patient H69 by Vanessa Potter
Catching Breath by Kathryn Lougheed
PIG/PORK by Pía Spry-Marqués
The Planet Factory by Elizabeth Tasker
Wonders Beyond Numbers by Johnny Ball
Immune by Catherine Carver
I, Mammal by Liam Drew
Reinventing the Wheel by Bronwen and Francis Percival
Making the Monster by Kathryn Harkup
Best Before by Nicola Temple

CATCHING STARDUST

COMETS, ASTEROIDS AND THE BIRTH OF THE SOLAR SYSTEM

Natalie Starkey

BLOOMSBURY SIGMA
LONDON • OXFORD • NEW YORK • NEW DELHI • SYDNEY

Bloomsbury Sigma
Bloomsbury Publishing Plc

50 Bedford Square
London
WC1B 3DP
UK

1385 Broadway
New York
NY 10018
USA

First published in the United Kingdom in 2018

Photo credits (t = top, b = bottom, l = left, r = right, c = centre)
Colour section: P. 1: David Blanchflower (t); ALMA (ESO/NAOJ/NRAO) (b). P. 2: Antarctic Search for Meteorites Program/Dr Katherine Joy (t); NASA (c, b). P. 3: NASA/JPL-Caltech (t); Open University/Dr Natalie Starkey (c); STFC (b). P. 4: NASA (t, b); Dr Natalie Starkey (c). P. 5: NASA/JSC (t) Dr Natalie Starkey (c); NASA/JPL-Caltech (b). P. 6: ESA/ATG medialab; Comet image: ESA/Rosetta/NavCam (t); NASA/JPL-Caltech/UMD (c); ESA/Rosetta/MPS for OSIRIS Team MPS/UPD/LAM/IAA/SSO/INTA/UPM /DASP/IDA; Context: ESA/Rosetta/NavCam (b). P. 7: ESA/Rosetta/NAVCAM (t); ESA/Rosetta/MPS for OSIRIS Team MPS/UPD/LAM/IAA/SSO/INTA/UPM/DASP/IDA (c); NASA/JPL (b). P. 8: Open University/Dr Jean-David Bodenan (t); Svend Buhl (c, b).

A catalogue record for this book is available from the British Library

Library of Congress Cataloguing-in-Publication data has been applied for

ISBN: HB: 978-1-4729-4400-9
TPB: 978-1-4729-5501-2
eBook: 978-1-4729-4403-0

2 4 6 8 10 9 7 5 3 1

Illustrations by Marc Dando

Bloomsbury Sigma, Book Thirty-three

Typeset by Deanta Global Publishing Services, Chennai, India
Printed and bound in Great Britain by CPI Group (UK) Ltd. Croydon, CR0 4YY

For Chloe

Contents

Preface

In school, our history lessons tend to focus on the past 6,000 years, the time encompassing human civilisation, even though people have roamed the planet for at least 200,000 years. *Homo sapiens* are an interesting species and we've achieved a great deal in our short time here on planet Earth. The past few generations have been particularly productive, we've even managed to blast into space to explore the Solar System surrounding our precious planet. The downside to being a human is our rather short existence on Earth; we can maybe hope to reach 100 years old, but probably not much more. This means that we aren't particularly good at contemplating the vast timeframes that the Solar System deals in: blocks of thousands, millions and billions of years. The Earth was born around 4.5 billion years ago – that's 4,500 million years, just a 'little' after (in geological timescales) the formation of the Solar System itself, which was 4.6 billion years ago.

Even if we consider the past 200,000 years of human existence – a period of time that seems hard to fathom in itself – it is an incredibly short interval compared with the age of the Earth. Using the age-old analogy of a 24-hour clock that started ticking when the Earth formed and which reached midnight at the present day, it would show that humans only arrived at a few minutes before midnight. Most of those 24 hours passed prior to the appearance of humans, and the planet achieved a lot in that time. For starters, the Earth had to form from a cloud of dust and gas and establish itself as one of the most important objects in the Solar System, one of the eight planetary bodies that owned its orbit around the Sun. It then had to create oceans and an atmosphere, and allow lifeforms to grow and thrive on and in it. Earth even had to recover many times from space objects repeatedly impacting its surface; it formed its own Moon; and it found a way to continually change its external appearance, destroying and re-forming its surface many times over, something that it continues to do at the present day, even if it isn't very obvious on the scale of human lifetimes.

As a teenager, I discovered that it was the subject of geology that allowed me to study the formation and evolution of our fascinating planet. Although history lessons were interesting – learning about humans and all they had achieved in recent years, whether good or bad – I liked the fact that studying geology, or Earth history as I see it, allowed me to delve much further back in time. Geology can let us learn how to form a planet and make it into an active, functioning, life-giving ball of rock. While studying geology I soon learnt to 'read' the landforms that made up the countryside, imagining how oceans once lay in places that were now positioned well above sea level, and how quiet mountains that were once volcanoes had spewed out lava and created all manner of intriguing landforms. Most importantly, I learnt to pick up rocks and look at them carefully to work out what they could tell me about the history of the planet they had been a part of. Later, by analysing these rocks I've been able to find out what their chemistry can tell me about the environment where they formed, and what this reveals about the formation of the Earth itself.

However, geology also lets us dig into the history of the other planets that surround us, and even the small space objects such as the comets and asteroids that orbit the Sun, too. After all, these celestial bodies are simply made of rock, ice and gases; they contain virtually the same mix of elements and rock minerals that we find here on Earth, having been born from the same cloud of dust and gas in interstellar space. The beauty of the comets and asteroids, in particular, is that they were the first celestial objects to form and therefore have a lot to tell us about the very earliest times in the Solar System, before and during the formation of our planet. If we want to understand where the Earth came from, and how humans eventually managed to thrive on this apparently important 'third rock from the Sun', then we must persuade the comets and asteroids to reveal their 4.6-billion-year-old secrets.

Just like time, it is also hard for most of us to fathom the immensely large distances that make up the Solar System, let alone the Universe. The distance to even our closest neighbour and only satellite, the Moon, is on average around 385,000km (240,000 miles): the same as travelling around the globe nearly 10 times! To make the Solar System easier to picture, I like to think of it as a city, with the different parts of it as neighbourhoods. The planets, comets and asteroids are all part of the overall city and its suburbs in some way, but the places where they are found represent very different areas of it, from the busy and lively downtown communities (*i.e.* the planets of the inner Solar System) to the quiet, calm and more sparsely populated suburbs (*i.e.* the comets).

The metaphor suggested here is one that we'll explore further in this book and I hope that it is useful for helping to examine the history of the comets and asteroids, delving into their family origins and learning about the places they came from and the neighbourhoods where they now thrive. We will see how they were built up from the essence of space itself and, as such, why they also play such an important role in human history. Without understanding the basic building blocks of the Solar System, the comets and asteroids, we can't begin to comprehend how the planets, and everything they contain, were formed.

We will also explore some of the groundbreaking missions that have encountered some of these fascinating space rocks, namely the NASA *Stardust* and European Space Agency (ESA) *Rosetta* missions. Space missions have marked an important turning point in our knowledge and understanding of these rocky and icy objects in our Solar System, and their findings will form the basis for any future exploration of comets and asteroids, whether that be for purely scientific study or for commercial gain, such as asteroid mining. However, despite the 4.6 billion years of history contained within the comets and asteroids, we shouldn't only look to the past, because these objects will play an important role in our future, or rather the future of our descendants; whether that be with the potential for them to destroy life on Earth following a collision, or saving life on Earth by providing us with vital resources that we might have depleted on our own planet. The only way that comets and asteroids can be of use to the Earth and its future inhabitants is if we study them to learn what they are made of and how they behave. Only then can we predict what they will do in the future, while giving us the opportunity to benefit from them, too.

CHAPTER 1

Introduction

My love for and interest [illegible] by a fascination with [illegible] fact that they can just loo[illegible] there, often looming over [illegible] people going about their [illegible] thought to the potential [illegible] those volcanoes that aren't [illegible] out stealthily beneath their [illegible] to unleash a sheer explosi[illegible] without warning. I think [illegible] Jekyll and Hyde – one minute [illegible] Luckily, the more that volcanoes [illegible] confident scientists can be in [illegible] in the process, saving [illegible] modern-day Pompeii [illegible]

It might seem like [illegible] [illegible] to comets and [illegible] [illegible] here my scient[illegible] the similarities between [illegible] they are all made of rock, [illegible] and they are more parts of [illegible] system relatively quietly [illegible] making their presence [illegible] are like the dormant volca[illegible] rather than dead, never [illegible] are so many comets and ast[illegible] vast majority of them, and [illegible] of others. Such is their [illegible] will probably never see [illegible] for as long as they might [illegible] this is where the Jekyll [illegible] asteroids becomes apparent [illegible]

CHAPTER 1

Introduction

My love for and interest in science was initially sparked by a fascination with volcanoes. I was allured by the fact that they can just look like mountains, quietly sitting there, often looming over large cities inhabited by millions of people going about their daily lives and not giving a second thought to the peaceful mound of rock nearby. However, those volcanoes that aren't extinct and have a great deal going on stealthily beneath their calm exterior, have the potential to unleash a sheer explosive, life-destroying power seemingly without warning. I think of an active volcano as a rocky Jekyll and Hyde – one minute so calm, and the next so angry. Luckily, the more that volcanoes are studied, the more confident scientists can be in predicting their behaviour and, in the process, saving the nearby population from a modern-day Pompeii.

It might seem like a stretch of the imagination to relate volcanoes to comets and asteroids, but as a space geologist, which is where my science career has eventually led me, I see the similarities between them all too clearly. To start with, they are all made of rock, but that is not where their similarities end. For the most part, comets and asteroids exist in our Solar System relatively quietly, moving serenely around without making their presence very obvious to us. In some ways, they are like the dormant volcanoes, those that are simply asleep rather than dead, never bothering us in any way. In fact, there are so many comets and asteroids out there that we can't see the vast majority of them, and we can only predict the existence of others. Such is their huge distance from Earth that humans will probably never see some of these far-flung space objects for as long as they might exist on the planet. However – and this is where the Jekyll and Hyde nature of comets and asteroids becomes apparent – if a comet or asteroid were heading on a collision course with our beautiful planet, even if it were modest in size, say around 400m (0.25 miles) across, it would have the potential to unleash all manner of destruction on Earth. A comet or asteroid impact could wipe out all of our planet's lifeforms. It might sound dramatic, but similar events have happened before. Scientists think that the demise of the dinosaurs was caused by a large meteor strike that changed our planet's atmosphere forever, throwing up huge amounts of debris when it impacted Earth's surface with such immense force that it would have melted the bedrock. The fragile ecosystems of our planet couldn't cope with this colossal shock any better today than they did in the era of the dinosaurs. Day becomes night, causing what scientists refer to as a 'nuclear winter'. Although the dinosaurs were the largest group of animals to go extinct 65 million years ago, the impact and ensuing chaos eventually led to the extinction of around 80 per cent of all animals living on Earth at the time.

We take for granted that our daily lives go on as they do thanks to the rotation of our planet and its journey around the Sun, meaning we experience day and night, and seasons that provide the more fortunate of us with plentiful food and resources to survive. Just like those city inhabitants living under that looming volcano and seemingly unaware of its quiet capacity for destruction, humans all over the planet live in blissful ignorance of the deadly potential for the mostly invisible, but possibly violent, comets and asteroids.

Maybe I've started to worry you, but remember, I'm a geologist and so the timescales I work with are somewhat longer than the ones normal humans contemplate. I'm used to thinking in blocks of millions, nay billions, of years to match the slow action of geological processes. However, you can breathe a sigh of relief because scientists currently predict that the Earth is unlikely to experience a major meteor strike in the next 100 years or so. Unless there are some major medical advances soon that will allow humans to live a lot longer than they are currently able, then we're all safe. Nevertheless, such predictions are based on the space objects that we know exist, the ones that we can see, measure and predict their orbital courses.

Unfortunately, there are some comets and asteroids that will be on Earth-crossing orbits in the future, possibly even within our lifetimes, that we can't yet see – the known unknowns, shall we say? These objects, if not spotted soon enough, will give us little or no time to react to impending annihilation. We might not have time to prepare ourselves for an impact, or to do something to prevent it from occurring. What we can be certain of is that, even if we are safe for now, it's very likely that our descendants are going to have to deal with the possibility that a comet or asteroid is heading for them.

In the long history of our planet, there have been hundreds of thousands of comet and asteroid impacts. Although the rate of impacts has slowed considerably since the first few million years of Solar System history – partly because the planets, asteroids and comets gradually settled down into their comfortable orbits – more are still expected to occur in the future. It's Solar System pinball, except we're dealing with lots of balls flying about all at once with a certainty that they will, at some point, collide with other objects in the game. It may sound like the stuff of movies and, of course, similar scenarios have already featured in a few Hollywood blockbusters, such as *Armageddon*. While sending oil drillers to an asteroid to break it up before it impacts Earth (the plot to *Armageddon*) may seem a bit far-fetched, there are already similar plans in place now, albeit using robots instead of the likes of Bruce Willis and Ben Affleck. In fact, breaking up potentially Earth-crossing space objects before they meet us is just one such method being discussed by scientists to prevent future impacts. We'll learn more about this in Chapter 10.

This brings us back to the volcanoes. As I said, scientists study volcanoes to predict their future behaviour and this has started to pay off in recent decades with some accurate forecasts of imminent eruptions resulting in the saving of many lives. So, just like the volcanoes, we must study comets and asteroids to predict their future orbit, to know if one is going to collide with us. But that is not all. We must also understand what they're made of, how heavy they are, how cold they are and how well they are held together, so that if one is heading for us we might be able to do something about it, perhaps by pushing it onto a new orbit or breaking it up. The main problem is, whereas it's relatively easy to go and visit a volcano – even an active one if you have the nerve – to poke it, sample its rocks and measure its gases to try to understand what makes it tick, it is far harder to do the same with objects in space. Travelling to space remains one of the biggest scientific and technological challenges humans currently face, even if only using robotic spacecraft. However, we absolutely must find a way to analyse and sample these space rocks to ensure a long and healthy existence for humans on Earth.

Visiting and sampling space

Leaving the safety of planet Earth to visit space – an inhospitable environment to humans – is hard, to say the least. Although vast numbers of spacecraft have left the surface of planet Earth over the course of the past 60 years, we've still barely explored even the parts of the Solar System closest to us, let alone the galaxy or Universe. The vastness of space means that we've had to be clever about the space destinations we've visited, initially choosing those nearest to us. Humans themselves have literally only touched the surface of Earth's fellow space citizen, the Moon. Of course, we've performed fly-by missions of objects in our Solar System, using robotic spacecraft to peep at the surfaces of many: the planets, the Sun, some comets and asteroids, and even far-away Pluto. We've caught glimpses, often at great distance, but sometimes at just a few kilometres above the surface, of these other worlds and have even performed basic analyses to determine what their surfaces are made of. When it comes to landing spacecraft on foreign worlds, however, robots have played a major role, with the missions to Mars being obvious examples.

Returning pieces of foreign worlds to Earth for our delectation is still not a common occurrence, despite the many space missions that were launched in the past century, and wherever they went. I don't mean to play down the amazing research and findings made by space missions that haven't returned samples to Earth. But, in order to make the firmest conclusions about what we think the planets, comets and asteroids are made of, and how they formed and evolved over the course of more than 4.6 billion years of Solar System history, we need to obtain direct samples from them that we can analyse with scientific instruments on Earth.

There is a comparatively large inventory of Moon rock on Earth, returned from the various lunar missions over the years. Apart from these samples, however, scientists have only collected specks of dust from anything else in the Solar System. Not that the precious extra-terrestrial dust samples are to be scoffed at, but their small size makes them extremely challenging to work with. The many kilograms of rock collected from the surface of the Moon are still being analysed in minute detail almost half a century after they were collected. However, despite the very small size of some of the other space rock samples, scientists have gleaned a great deal of information about the objects from which they originated, and still have material left to analyse. Such analyses have revealed key details about the formation of the Solar System itself, often using just picograms (one trillionth of a gram) of rock dust.

Even with the huge challenges that space exploration presents, and our limited time observing, and roaming on, foreign space objects, scientists have been able to start building up a detailed understanding of the history of some of these alien environments. They have started to find out how they formed, what they're made of and whether they contain, or have ever contained, life. Scientists have learnt a great deal about some Solar System objects even if they've only obtained samples from a few choice locations on their surfaces, then extrapolated the data and interpretations to the parts they haven't visited. Of course, they can't be 100 per cent certain their stories are correct, but that is part of the scientific process; ideas are always evolving as newer information is obtained. Maybe if aliens were parachuted deep into the Sahara or Atacama deserts on Earth they might deduce that the entire planet is devoid of water and life and wonder how anything survives here. They might have seen our blue oceans as they came careering towards Earth, but unless they had planned for some serious travel on our planet once they arrived then they might need a second mission to discover the all-important moisture that keeps it going. Such an analogy suggests that there is still much to learn about our planetary neighbours, even those we've visited many times, and our current understanding of how these environments formed may change as we explore our Solar System further and in more detail than has been possible to date.

Discovering the billion-year-old secrets of our Solar System has, and still is, an incremental process where we must progressively work on more and more complicated missions, trying to travel further, stay longer or deliver more complex instruments to far-flung destinations. In recent years, we have entered a new space race, where private companies are increasingly investing in pushing the boundaries of what we think is possible in terms of mission timelines and achievements, and heralding a new era of space exploration. The reason? There is the potential for great wealth to be made in space. The much-touted plans for asteroid mining that not long ago might have seemed like science fiction are starting to become reality and that is thanks, in part, to the commercialisation of space. Governments and their space agencies are, in many cases, working with commercial would-be space miners to share their experiences while gaining something for themselves in return, such as access to launch vehicles. Without the desire to make money from space, we wouldn't be progressing towards the goal of exploring other planets, such as Mars, or making the red planet or the Moon a human base, as quickly as we are now.

Part of the problem with space is the obvious issue of the vast distances to even our closest Solar System neighbours: distances so immense that it can take more than a human lifetime, as well as a huge amount of energy and years of complicated planning, to journey across. Although space mining may first concentrate on some of the closer objects to Earth, such as the asteroids, it must be remembered that these are still further away than the Moon. To compound this issue, landing a laboratory space robot, let alone a human, on a distant object of which we know very little and of which we may only have a blurry pixelated image is no simple task. It might even seem a bit crazy. How can you plan to land on something when you don't know how hard or soft its surface is, what it's made of, or what shape or exact size it is? Despite the unknowns, this seemingly impossible task is something that has been achieved by one particular space mission in recent years, the ESA *Rosetta* mission, proving that scientists are capable of achieving amazing things that most people would consider impossible at first sight. But more on that in Chapter 8.

Comets and asteroids

The size of the Solar System, and the problems in visiting its farthest corners, mean that scientists are still drawing a detailed

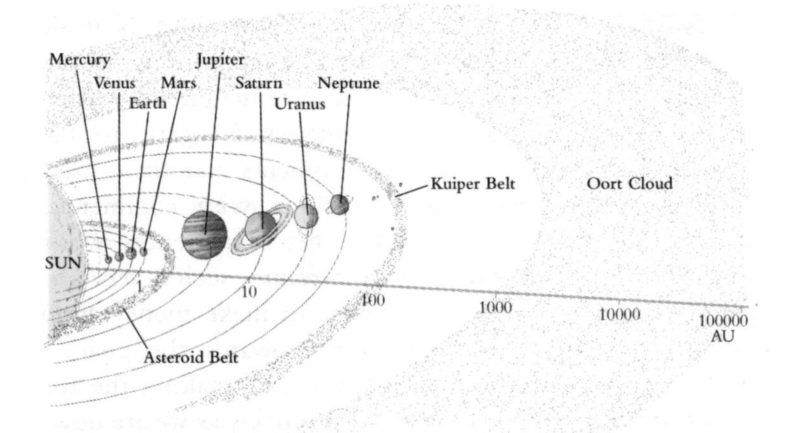

Figure 1 *The objects in the Solar System. A schematic illustration to show the relative positions of the eight planets, the Asteroid Belt, Kuiper Belt and Oort Cloud, including a scale in Astronomical Units (AU).*

picture of what the real estate surrounding our average star contains, and trying to understand how it formed. Our Solar System has a long, dramatic and fruitful history, and it will take many more years for us to fully appreciate the complexities of how it was built, especially if we are to comprehend the tiniest and farthest objects out there. Some of these are the icy comets and rocky asteroids that make only occasional and fleeting visits to the inner Solar System, when they whizz past the Earth at great speed. To understand comets and asteroids, and their importance for realising our place as humans in the Solar System, we must visit them, sample them and analyse them to find out what they contain, what they look like and how they behave. This means that we either have to go to them in their natural habitat, in some cases at the very edge of our Solar System, or we have to catch up with them when they enter the inner Solar System.

When comets and asteroids make their way into the inner Solar System it is because they have been diverted from their normal orbit – which usually keeps them very far away from us – and in towards the Sun because of gravitational interactions with large planets. Comets and asteroids can be viewed as visitors from a distant place, not only in space but also in time, as they bring with them material collected up from the very beginning of the Solar System – in the case of comets, 4.6-billion-year-old gases, dust and ices. Hence, comets are sometimes called 'dirty snowballs'. If scientists can sample the material contained in these ancient space objects, it effectively allows them the chance to travel back to the very earliest days of the Solar System. Without the invention of a time-travel machine, this is our best chance of understanding a crucial time in our history, the birth of the Solar System and everything contained within it, including ourselves.

Before the planets formed, the Solar System was nothing but a swirling cloud of gas and dust, which was itself travelling through space. If we want to draw on the Solar-System-as-a-city metaphor here, we can think of this as the peaceful and luscious green countryside existing before our metropolis was built. This swirling cloud eventually became the construction site for everything we see today, including the planets, asteroids and comets, from the largest gas giant, Jupiter, to the smallest specks of dust that travel around in the space between the planets.

The outer edge of this cloud eventually became the immensely cold and far fringes of today's Solar System – places that are hard to see from Earth with even the most powerful telescopes, where barely any of the Sun's energy can reach. This represents the true outskirts of our city, the boroughs that barely feel like they are part of the conurbation at all. Many comets formed at the very edge of this cloud and for well over 4 billion years they have lingered in this remote neighbourhood. These comets are so far away that they could be represented by the rural communities that often surround large cities. In the same way that daily life in these farming regions is little affected by the hustle and bustle of the capital, these comets have been able to remain perfectly in deep freeze, preserving the early-captured planet-building ingredients that they scooped up billions of years ago. This is one of the reasons why comets are so interesting. They contain the starting ingredients for the Solar System and have the potential to reveal the secrets about how it formed. In this respect, comets can shed light on the reason for the existence of humans here on Earth.

Asteroids, on the other hand, can be thought of as the rockier counterparts of the icy comets. Instead of forming in the cold outer reaches of the early cloud, they formed much closer to the young Sun, where temperatures were, unsurprisingly, hotter. The asteroids formed slightly later than the comets, but earlier than, or concurrent with, the planets. Just like comets, asteroids allow us to glimpse into an earlier time in Solar System history, one to which we would otherwise have no access, as they record information about the important phase of planet building. We can think of asteroids as the leftover Solar System building rubble that wasn't incorporated into one of the major city skyscrapers, or planets in the case of the Solar System. The asteroids missed out by not being in the right place to be absorbed into one of the growing balls of rock that became Mercury, Venus,

Earth and Mars. The benefit for the asteroids that escaped being gobbled up by a greedy planet is that their evolution was all but stopped just after their formation, meaning they preserve key information about an important phase of Solar System formation.

The planets, although made of the same stuff – hardy rocks and metals – continued to evolve over the intervening billions of years. In fact, they still do, particularly in the case of a planet like Earth that is geologically active, constantly moving its outermost layers of rock around, slowly destroying and forming new landmasses. Studying asteroids allows us to understand the important planet-building phase of Solar System formation without having to unravel the complex effects of over 4.5 billion years of geological activity that is required when we study the planets. Knowing what asteroids are made of means we can understand how our planets were put together and what they contain.

Life on Earth

As humans, we may have a lot to thank comets and asteroids for. Our very existence on Earth's surface is likely to be because, long ago in our planet's history, these cosmic rocky and icy objects bombarded the Earth's surface. With them, scientists think they might have delivered the very ingredients that allowed life to take hold – including the all-important life-giving water – to our barren lands. It is certain that space objects impacted the Earth's surface billions of years ago, evidenced by the cratered surface of the Moon that starkly displays this early frenzied phase of Solar System history. Such evidence has been all but lost on Earth because of the geological action of plate tectonic activity: movements of the broken pieces of the Earth's carapace – its rocky crust and some of its underlying slightly squidgy rock mantle – that bring about the destruction and renovation of its surface features. However, life hasn't been found on a comet or asteroid yet, and is highly unlikely to be, so why is it thought that these small space objects might be the life-givers? Well, what has been found on both comets and asteroids is organic matter, including amino acids – compounds that form the basic building blocks of proteins that are responsible for life – and, of course, plenty of water. It's not that these cosmic rocks delivered life as we now know it – complex organisms that can see, hear, think and feel – but they might have delivered life's all-important building blocks: proteins, amino acids and the essential solvent, water, needed for life to thrive. All that these life-giving ingredients desired was a calm, welcoming environment to bed down in; somewhere not too hot and not too cold. Earth, being located at just the right distance from our average star, the Sun, was apparently perfectly placed to provide the ideal conditions for life.

Without visiting comets and asteroids we will find it hard to truly assess whether life on Earth, or the precursors for it, hurtled through space to get here. The fact that some of the basic building blocks for life have been discovered on these objects in space is one thing, but finding a link between these extra-terrestrial ingredients and the advanced lifeforms on Earth requires much further investigation. For instance, one of the areas that must be explored are the amino acids. These organic compounds come in two varieties: so-called left- and right-handed. The interesting thing is that the amino acids that build the proteins for life on Earth rely solely on the left-handed variety and never their mirror-image counterparts, the right-handed version. Scientists must find out whether the same is true for the amino acids found in space. If the extra-terrestrial amino acids are found to be exclusively right-handed, then science has a slight problem in explaining how they could be responsible for creating humans, and all the other life on Earth. This is something that we will explore further in Chapter 5.

Another aspect of this story that needs further examination is the composition of water in space compared with Earth. Although measurements of water in various space objects have now been made, the picture that is being built is not as simple as first predicted. In essence, scientists would like to see if the water on comets and asteroids has the same 'flavour' as the water on Earth. If none of the water in space is found to match the exact concoction of that on Earth, it might mean our planet began life with all of its water in place from the outset. However, this is a theory that many scientists believe to be unlikely, even if it hasn't been disproven yet. The reason for this is that, early on, Earth was a very inhospitable place. For one, it was a very hot, molten ball of rock. It would have struggled to amicably host even the simplest starting blocks of life – let alone a very volatile species such as water – for at least the first few million years of its existence. So, understanding whether Earth's water was delivered from outer space also impacts the likelihood of organic material being parachuted onto our surface in the same way. If no asteroid or comet contains water that matches Earth's, then scientists need to rethink how life could have taken hold on our planet, too. Currently, these are research questions that remain unanswered and are some of the reasons scientists are so keen to visit and sample more of these fascinating space objects.

Visiting comets and asteroids

One of the problems of sampling and measuring comets and asteroids is that they reside very, very far away from Earth. In fact, much like the city dweller without access to a car who never ventures out as far as the greenbelt, no spacecraft has ever travelled as far as the outermost comet neighbourhood. Luckily, comets and asteroids occasionally exit their cold, dark home and pay a visit to the inner Solar System. We can, perhaps, think of these objects as the out-of-towners travelling into the city for some weekend sightseeing. These people zoom through the outskirts of the town on their high-speed train line, just like the comets and asteroids that whizz past the planets on their orbits around the Sun. When these usually far-flung space objects make such journeys, it gives scientists a fighting chance of getting near enough to them, within a reasonable timeframe, to study them up close. Sometimes this even provides scientists with precious samples that are useful to understand what the comets and asteroids are made of. This has only been achieved a handful of times, and in only two of these cases have samples been returned to Earth: with the *Stardust* and *Hayabusa* missions. Excursions taken by comets and asteroids from the everyday orbits in which they may have spent billions of years also allow them to be studied more easily with telescopes. This means scientists can learn a great deal about what they look like, how they behave and even what they contain.

Of the space missions that have left our planet and ventured out to visit some nearby comets and asteroids, one of the most recent, and arguably the one that has most advanced our understanding of icy comets, is the *Rosetta* mission. Launched in 2004, *Rosetta* didn't return samples to Earth but achieved a great deal nonetheless, by catching up with a comet as it whizzed through the inner Solar System, and entering into orbit around it at very close proximity, beaming back awe-inspiring images of the comet's surface. This plucky little space mission even managed to land a science laboratory on the comet, one of many magnificent firsts achieved by the mission. Although *Rosetta*'s target comet was one of those that had exited its cometary home to journey into the inner Solar System, it didn't necessarily make it any technologically easier to get to – just a bit quicker perhaps.

The mission took over 10 years to plan, requiring an enormous amount of technical expertise, and it had a further 10-year journey through space after launch just to match the speed of the dirty snowball it was chasing in exactly the right location in space. This journey involved zooming back past Earth and Mars to gain gravitational energy, and a transit through deep space, a mission in itself even before it had got anywhere near the comet. Even though *Rosetta* wasn't designed to return samples to Earth, the laboratory it delivered to the comet's surface performed real scientific experiments and the orbiter had its own suite of instruments working away throughout the mission, too. The flood of data beamed back by *Rosetta* is still being interpreted, and will continue to teach us a huge amount about this comet, with the strong possibility that many of the findings can be applied to other comets not yet visited. Scientists have already learnt what it's made of, how its surface changes with each orbit of the Sun and how it reacts to its encounter near the Sun, but one of the most interesting findings so far concerns the comet's water. These findings will be covered in detail in Chapter 8; all I will say for now is that the story of water and life on Earth is still being written.

Despite the fact that we may have to thank comets and asteroids for our very existence here on Earth, there are some serious downsides to them entering the city centre of the Solar System. Just as they did billions of years ago, these small, high-speed space objects still have the potential to collide with us today. Whereas once they might have brought life to Earth, now they have the capability to destroy it, too! Luckily, our Solar System has calmed down a bit in the last three or so billion years, meaning that large comet and asteroid collisions with planets are now much less common. But they still can, and will, occur in the future. A huge number of space rocks collide with the Earth every year but most of them are very small, mere dust-sized rock fragments that rain down imperceptibly. (We'll learn more about these in Chapter 4.) However, some of the space rocks that collide with us are quite large, producing fireballs as they encounter the Earth's atmosphere, and some that have occurred in recent centuries were big enough to cause damage on the ground, destroying buildings.

Comets and asteroids in history

Historically, comets have been greatly feared, probably because the human psyche has known that the relationship between the Earth and small space objects that orbit close by is a rather delicate and potentially disastrous one. Halley's Comet, for example, is one of those that has caused people the most concern over the years. Yet it is also one of the more famous comets of our time thanks to its brightness, being seen regularly as it makes a sojourn near the Earth every 75 to 76 years: a 'short-period' comet. Halley is responsible for creating the stunning Orionid and Eta Aquarids meteor showers from the immense trail of dust it leaves in its wake as it makes its way through the inner Solar System. Nonetheless, the regular appearance of Halley has long been seen as an evil omen on Earth. One of its most notable appearances was in 1066, as depicted in the Bayeux Tapestry. At that time, Halley's Comet was thought to have passed very close to Earth, appearing large and bright in the sky which, despite being an awe-inspiring sight, with a lack of understanding may also have been alarming. The year of 1066 is also renowned for the Norman invasion of England and the resultant Battle of Hastings. In the tapestry, King Harold II of England is portrayed during his coronation and Halley's Comet appears on the far right surrounded by apparently terrified people. Harold is being told of the comet's appearance, and a fleet of invading ships can be seen in the tapestry's lower border, the ships that make up the fleet for the invasion by William the Conqueror. This suggests that the tapestry makers saw the comet as a bad omen for Harold. Later that same year Harold was killed in battle. However, for William the Conqueror who defeated Harold, Halley's Comet was evidently a good omen. So, whether comets are seen as 'Jekyll' or 'Hyde' is probably open to personal experience.

Going even further back in human history, there is evidence to suggest that the Younger Dryas event – a 'mini ice age' on Earth that might be linked to the extinction of the woolly mammoth – was caused by a comet strike. For years, some researchers have suggested that an impacting cosmic object was responsible for throwing the Earth into a cold snap that lasted 1,500 years. The timing of this event coincides with the Earth emerging from a major ice age, when global temperatures were on a generally increasing trend, and when humans were spreading far and wide to make the most of the new warmer climate. However, there is much debate in the research community about this impact, partly because no craters have been found. There is still no consensus on what else could have caused the Earth to experience such a sharp climate reversal.

Intriguingly, researchers have recently interpreted ancient stone carvings that appear on the so-called 'Vulture Stone' at Göbekli Tepe in Turkey – possibly the world's oldest temple site – as depicting the constellations and a swarm of comet fragments striking the Earth. They calculated where the constellations would have appeared above Turkey in the past to conclude that the comet strike portrayed in the carvings probably occurred at 10,950 BC, coinciding with the start of the Younger Dryas as recorded by ice cores from Greenland. The possible coincidence in timing between these events is interesting, but since it is not proven at this stage we tend to keep an open mind, as there may be a different explanation for the sudden drop in global temperatures. However, if a comet striking the Earth really was responsible for the Younger Dryas, then we might have a lot to thank the comet for. The difficult climate conditions that occurred around this time forced communities on Earth to work together to find new ways of growing and maintaining crops such as barley and wheat, where previously the warming climate had allowed wild crops to grow freely. As such, it could be argued that a comet strike was responsible for the emergence of farming and civilisations, as people established themselves together to grow food. If this can be proven, then it is clear that comets have played an important role in human existence: a big claim and one that is very exciting.

During the past 200 years – despite there being a period of fast-developing scientific enlightenment – so little was known about comets that people were still fearful of their appearance. In 1910, mass hysteria broke out in New York City when people feared they would be killed by a passing comet: the infamous Halley once again! Using the first spectroscopic data to be collected on a comet, astronomer Camille Flammarion discovered that Halley's tail contained a gas called cyanogen. The fear was that, as Halley passed the Earth, this gas would react with nitrogen in the Earth's atmosphere to create laughing gas. Flammarion even suggested that the cyanogen might permeate the entire planet and destroy life. The level of hysteria surrounding these predictions was so severe that 'comet pills' with leather inhalers went on sale, along with 'anti-comet umbrellas' to protect people from the comet as it passed by the Earth. Of course, nothing of the sort happened. The cyanogen gas, although present in Halley's tail, was at such a low concentration that Halley passed by serenely, continuing its peaceful elliptical orbit of the Solar System once again. Halley next passed the Earth in 1986 and became the first comet to be observed in detail by spacecraft: namely the ESA *Giotto* and Soviet *Vega*. The next opportunity to observe Halley will be in 2061 and, as the most active of the short-period comets, there will once again be plans to observe it in detail, as it is such a good opportunity to get close to a space object that is usually so far away.

Despite their shared potential for disaster, it seems that asteroids have generated less fear over the years than comets. The reason for this might be that, unlike comets, they tend not to be 'active' as they enter the inner Solar System, making them harder to spot in the sky. As comets and asteroids come close to the Sun they are heated up. A comet can produce a tail of comet dust and gas for thousands of kilometres in its wake, and in some cases, this magnificent tail can be seen with the naked eye. However, asteroids generally don't exhibit such behaviour because they don't usually contain volatile materials that react to the Sun's heat. This means that unless you have access to a highly specialised telescope, you're unlikely to be aware of an asteroid passing by. However, the fact that asteroids can lurk inconspicuously out there doesn't necessarily make them any less dangerous.

We must remember that during the history of the Earth, space objects just like Halley have impacted our planet's surface on many occasions. We can think of these impacting rocks as a double-edged sword; they may have delivered water and organic material to our barren planet, providing us with the components for life, but at another time they may have been responsible for devastatingly large impacts that destroyed any life that had grown up on the planet's surface.

So, the big question is, should comets and asteroids be feared or revered? It might be that our very existence here on Earth is only possible because of them, but at the same time, unless we understand them better, they could suddenly trigger the start of our demise. The more we know about these icy and rocky worlds, the better chance we have to predict if, and when, they will collide with us, and maybe even do something to mitigate their potentially disastrous effects.

If we want to remain in awe of these fascinating, small but important and powerful space rocks, it is paramount that we understand more about them, learning about their composition, structure and behaviour. Ultimately – and this is probably the most exciting aspect – the more we learn about comets and asteroids, the more we will understand where we came from, where everything in our Solar System came from, and how it was put together. These cosmic rocks really do have a lot to tell us.

CHAPTER 2

A 4.6-Billion-Year Journey into the Deep Freeze

To understand our place in the Solar System, we must start at the very beginning, at a time before any of the space objects we can see within it today had formed. The opening days of the Solar System are not based on geology, because at this time there were no rocks. In their place were all the

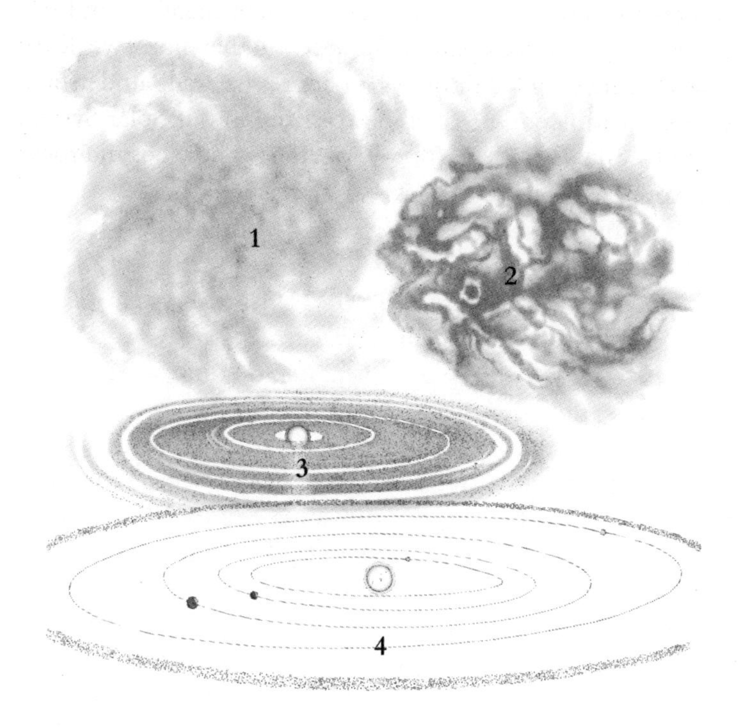

Figure 2 *The first few million years of Solar System formation. Beginning with a giant cloud of gas and dust (1), clumps begin to form that can turn into dense cores and the precursors to stars (2). The cores condense to form into young stars surrounded by dusty discs (3) from which planets can form and a Solar System is created (4).*

components of rocks – in fact, there were all the ingredients required to form an entire Solar System, including the ones necessary to create complex organisms like ourselves. To get to where we are today, these ingredients needed some processing. This was essentially mixing and baking at high temperatures, so we can think of the development of the Solar System as a sort of extreme cookery book.

Forming the Solar System out of the interstellar ether

Around 5 billion years ago our Solar System didn't exist. In its place was a region of interstellar space composed of a highly dispersed cloud of particles, so not very much at all. This expanse of space was in our galaxy, the Milky Way, which is just one of a mere 100 billion galaxies in the Universe. This is a mind-boggling number but fear not, we will focus on just one of the possible 400 billion stars contained within our galaxy – the Sun, which hosts our Solar System.

The Milky Way is not only made of stars – there are the remnants of dead stars that have completed their life cycle and also the interstellar medium (ISM). The ISM is essentially the space between stars, which doesn't necessarily sound very impressive and it is very diffuse, containing just one atom of matter per cubic centimetre. However, the ISM is very special indeed. It is the birthplace of new stars, being composed of the gas and dust required to make entire star systems. ISM gas is mostly hydrogen but there is also some helium left over from the Big Bang, as well as tiny amounts of heavier elements that were ejected from dying stars that have come and gone. The dust in the ISM doesn't look exactly like dust on Earth; it is composed of very small particles made of carbon, ice and silicates (compounds containing silicon and oxygen which constitute the largest group of rock-forming minerals). These gas and dust ingredients are literally the seeds of new worlds.

As the Milky Way rotated – something it's done around 60 times since it formed over 13 billion years ago – it mixed up the gases and pieces of dust contained within it, collecting them into vast molecular clouds. The increasing density of such clouds allows the elements that are present to appear in molecular form: hydrogen gas (H_2) instead of simple elemental H, for example. In fact, scientists have detected hundreds of different types of molecules in such clouds, including water, ammonia and even amino acids. Deep in the densest parts of these clouds of galaxy debris, new star and planetary systems can form. Sometimes the stars that are formed are just like our Sun; they have planets orbiting around them, and some of these planets might even be just like our Earth. Although scientists are yet to find any other stars with life on their accompanying planets, it doesn't mean it's not out there somewhere. They probably just need to keep looking, and so they are.

More than half the stars in the Milky Way are older than 4.6 billion years, meaning that they arose before our Solar System. Other stars are still forming, or are yet to form. Indeed, planetary systems are continually being built and eventually destroyed all over the Universe, in ours and all the other galaxies that exist. It's a never-ending process that constantly recycles old star material into new stars and planets. Unfortunately, we can't travel to any of these other stars yet, even the ones within our own Milky Way, as they are just too far away. What we can do is look out into the sky using powerful telescopes to capture images of these stars at different stages of their existence – including, in some cases, before and during the formation of their accompanying planets. The useful thing about looking at other star systems is that it gives scientists an idea of the processes that took place to form our own star and planets. In fact, it is thought that on a very broad scale the same processes are at work throughout the Universe to form every one of the amazing planet neighbourhoods that exist. The current model for this is the 'nebular hypothesis', which suggests that a star and its planets form from part of a swirling molecular cloud as described above. This is why molecular clouds are sometimes called 'stellar nurseries', because they have the potential to give birth to a whole crèche of baby stars.

The early forms of the nebular hypothesis were developed around the same time during the eighteenth century by the

German and French scholars Immanuel Kant and Pierre-Simon Laplace respectively. Kant and Laplace proposed slightly differing ideas, both theories following on from earlier concepts put forward by the Swedish scientist Emanuel Swedenborg, who suggested that the Solar System was formed from a nebula. Kant argued in his *Universal Natural History and Theory of the Heavens* (1755) that rotating gaseous clouds gradually collapse and flatten to produce stars and planets. Laplace argued in his *Exposition du Système du Monde* (*Exposition of the System of the World*), released in 1796, that the Sun was surrounded by a hot atmosphere that cooled, contracted and threw out material as it rotated, which condensed to form the planets. Development of these early models continued after the scientists' deaths, most notably by Soviet astronomer Viktor Safronov, who can be credited with the most modern version of the hypothesis widely accepted as the solar nebula disc model (SNDM), or nebular hypothesis. He developed his ideas in the mid-twentieth century and presented them in *Evolution of the Protoplanetary Cloud and Formation of the Earth and the Planets* (1972). The model was originally proposed to explain the origin of the Solar System, but has gone on to become a widely accepted view of how all star systems originated.

However, a hypothesis is only a prediction of what might have occurred and the only way to test whether it is correct is by gathering data and seeing how they fit the hypothesised model. Of course, telescopes can be used to look at other planetary systems and use what is seen to test and gradually refine the hypothesis. This, in turn, sheds light on how our own planets formed on a very broad scale, on the scale of planets. However, looking out into space doesn't really allow us to delve down to the very fine print of the formation of a planetary system, down to the level of the individual rocky grains that built up to form the larger planetary blocks. Unfortunately, we can't just observe other stars to understand the variability of the planets in the Solar System. In particular, if we want to comprehend how our own planet hosts so many varied life forms, it might not be helpful to look at other stars if none of their accompanying planets host life. If Earth is the only place in the Universe that we know has life, then we really must focus on the Solar System itself to appreciate how our beautiful planet, and those that surround us, formed.

So, the best way for us to truly grasp what has occurred over the course of the past 4.6 billion years in our little corner of the Milky Way is to study the Earth and the objects that formed around it – *i.e.* the planets, the comets and the asteroids. It is these sometimes rocky, sometimes icy and sometimes gassy objects that can help us understand how we progressed from a vast swirling cloud of gas and dust hanging out in a spiral arm of the Milky Way to the complicated architecture of objects rotating around the Sun that we see today, including at least one object that hosts complex, intelligent life.

Studying the planets themselves is all well and good, but they tend to make things rather complicated when attempting to unravel the timeline that can take us back 4.6 billion years to the beginning of the Solar System, when the first rocks were forming from the molecular cloud. The reason for this is that the planets have experienced vast changes in their long history since formation. They have been melted and re-formed many times over, experiencing violent impacts from space and then undergoing planetary-scale processing such as plate tectonics, of which we'll learn more later. In fact, because of these processes, the planets have almost completely hidden away or lost much of the information about the molecular cloud from which they formed. If scientists wanted to use the planets to understand the molecular cloud, then they'd have a great deal of geological unravelling to do. Luckily, this is where the comets and asteroids come into play, as not only were they formed of the exact same starting ingredients as the planets but they also formed before them, right after the Sun came into being. Many of the comets and asteroids have even remained the same over the course of their 4.6-billion-year existence, preserving vital information about the composition and structure of the molecular cloud. These rocky and icy objects are the Solar System foundations upon which everything else was built and they must be studied in order to probe back to the beginning of our time.

Before we look at comets and asteroids in detail we must understand a bit more about how exactly they were formed, so let's start at the very beginning, deep within the cloud of gas and dust molecules that was floating through space in the Milky Way. Something happened around 4.6 billion years ago that caused the large molecular cloud to collapse in on itself. Scientists don't know for sure what that trigger was. It might have been the influence of a passing star or the shock of a supernova – a titanic explosion at the end of a large star's life – but whatever it was, it was big. What happened next is the key to our Solar System's existence in the Universe. As the cloud collapsed, it moved the matter within it, concentrating it into denser regions, forming clumps. The cloud began to break into smaller pieces as these dense clumps pulled in more and more matter until they reached stellar mass, the mass required to form a star. The dense clumps of matter began to rotate and their increasing interior pressure caused them to heat up, resulting in them undergoing gravitational collapse and the formation of a star.

Gravitational collapse, simply put, involves the gas and dust falling in towards the centre of the cloud, causing it to contract. This contraction causes the already rotating cloud to spin faster. In Carl Sagan's famous and brilliant book *Comet*, this process is likened to that of an ice skater performing a pirouette. As they bring their arms in closer to their body they spin faster: a lovely way to explain angular momentum, which is exactly what's at play here. As the particles move towards the centre of the cloud, they are slowed by friction and lose energy as heat. The result is that the centre, or core, of the cloud gets warmer and warmer … and warmer. We're talking extremely high temperatures. In fact, eventually the density and temperature in the centre of the cloud increase to such a point that thermonuclear fusion reactions are possible. This is what brings about the birth of the star or, in the case of our Solar System, the Sun. Whereas we began with a highly diffuse cloud of material with no focus, fairly quickly most of the mass has ended up in the centre, at the star. The remaining material is dispersed around it in a large cloud that will soon form the planets, asteroids and comets. At this stage of growth, the cloud is known as a solar nebula, where *sol* is Latin for 'Sun' and *nebula* is Latin for 'cloud'; thus, a cloud around the young Sun. Only about 10 per cent of the original molecular cloud material actually gets locked up in the stars and planets; the rest is literally blown away into the interstellar medium, where it is once again available to be recycled into new molecular clouds.

The dust and gas in the solar nebula surrounding the infant Sun continue to fall, under gravity, inwards to the centre and the whole system continues to spin. These processes can't continue forever, though, otherwise the nebula would just get smaller and smaller and eventually disappear in a black hole. The reason this doesn't happen is centrifugal force. This is a clever bit of physics that acts to balance the inward pull of gravity – stopping the gas and dust falling inwards – but also slowing down the spinning. Sagan likens this process to whirling a bucket of water on a rope around your head. The water doesn't spill if it's being whirled quickly enough, and that's because of centrifugal force. However, the centrifugal force only acts in the plane of rotation. This is exactly why the rather irregularly shaped cloud starts to flatten out to form a disc, because matter outside the plane of rotation continues to fall in. In space, this is what results in a protoplanetary disc – a rotating flat disc of dense gas and dust that is observed around many young stars. At the present day in our Solar System it is represented by the equatorial plane formed by the planets, which is called the ecliptic.

Forming comets and asteroids

Next, we need to understand how it is possible to form so many varied Solar System objects, such as planets, asteroids and comets, from a swirling disc of interstellar material. At the very outer fringes of the disc it was freezing cold because it was so far from the Sun. Cold enough, in fact, that even ices of methane and carbon monoxide were stable. The solar nebula gas and dust that were existing in this far-flung disc region condensed out to form 'fluffy agglomerates' of dust and ice – loosely packed arrangements of these solar nebula materials – that were then attracted to one another by small electrostatic and gravitational forces. I like to think of the fluffy, icy, dust balls as being almost like the little dust bunnies that exist on the floor of my house when I need to clean. They eventually start to find each other and grow into larger and larger dust bunnies the longer I ignore them. In the solar nebula, the fluffy agglomerates loosely stuck to each other, growing to form larger and larger objects, some of which became the size of present-day comet nuclei, the solid central part of a comet. The average comet nucleus is around 10km (6.2 miles) in diameter, but the largest ones can be up to hundreds of kilometres across. There are many much smaller, too: those dust bunnies who never found a mate.

These icy cometary space rocks are not very dense. In fact, most comets probably have densities just half that of water. The reason is twofold: water ice is less dense than liquid water – hence ice floats on water – and the comets contain a lot of empty space, so there are literally gaps inside of them. The reason for this is that collisions between the icy, dusty, fluffy agglomerates were infrequent (because the bunnies were very dispersed in the outer disc) and when collisions occurred they were at much slower speeds, at much lower impact (kinetic) energies. Think about rolling a snowball. The more you press the snow together, the harder and denser the snowball becomes as you compact the ice crystals. The more compacted your snowball, the more pain it will inflict on your target. A comet is not much different, except here you must combine some mud into the snowball, too. This helps to reiterate why comets can often be known as dirty snowballs – they are made of loosely packed dust and ice.

Comets are said to closely resemble the early solar nebula cloud from which they formed; they are essentially a scoop of cloud, and they are intrinsically very fragile in structure. It is for this reason that scientists find them so important to study when they want to know more about the early Solar System. Comets can be thought of as time capsules that have remained in cold storage in the place where they were formed over 4.6 billion years ago. This means that to this day, and despite all the time that has passed, comets are still simply fragile agglomerates of solar nebula dust and ice. When scientists get the chance to study and sample comets, they have direct and unique access to an ancient era of Solar System history. Without the comets, it would be extremely difficult to know what our early Solar System was made from, what its starting ingredients were.

The asteroids, on the other hand, do not have a structure that resembles a delicate scoop of the early solar nebula in the same way. This is because the asteroids formed much closer to the Sun, where the disc material was hotter than that in the outer disc. The high temperatures nearer to the Sun meant that ices of any element could not survive, leaving behind just gases, silicate rock dust and metallic elements. The original low-density, fluffy, solar nebula agglomerates that might have started to form before temperatures rapidly rose, were transformed into new materials by the heat: Solar System mineral phases and metals that were sturdy and dense. The asteroids grew up in much the same way as the comets, through gravitational attraction and impacts of initially small grains that gradually clumped together to form larger and larger rocks. The main difference between the asteroids and the comets is that the grains forming the asteroids were made of the denser rock and metal, and the comets were made of solar nebula fluff. When newly formed asteroids collided, whatever their size, not only were the collisions more frequent in the inner disc, but they were also more violent. In some cases, inner disc collisions were so ferocious that asteroids were completely disintegrated when they met one another. In other collisions, they could be combined and compacted together, leaving them with little pore space and making them much sturdier and denser space objects overall.

Forming planets

Over time, many of the small asteroids formed into more substantial objects, with some growing large enough to become planetesimals. These objects ranged from a few metres to hundreds of kilometres across and were positioned on much more random orbits than those of the planets today.

If the planetesimals were very big, then their own self-gravity would pull them into a roughly spherical shape, but smaller asteroids – lacking the required gravity – ended up being rather oddly shaped – broadly rounded but not spherical. The time period of the planetesimals was one of 'survival of the fittest'. They had to jostle for an orbit in space and cataclysmic collisions were still common that, in some cases, completely blew them to pieces. Eventually, the chief planetesimals gobbled up smaller ones as well as asteroids in their path, to grow into even larger objects. These became the rocky planets we see today. The planets established themselves as the leaders of the pack, gravitationally excavating their clean orbit around the Sun. From Jupiter outwards is the neighbourhood of the giant planets, grown from a region of the disc similar in temperature to the very outer disc where the comets reside. As we learnt, here it is cool enough for volatile icy compounds to remain solid. The ices in the vicinity of the newly forming giant planets were also present in higher quantities compared with the silicates and metals that formed the rocky inner planets. This meant the outer planets could grow massive enough to capture large atmospheres and become the especially vast Solar System bodies that they are today.

Some of the small rocky objects present in the early inner disc managed to escape being consumed by one of the growing inner planets. They remained as asteroids and mostly ended up convening in the main asteroid belt between Mars, the outermost rocky planet, and Jupiter, the innermost gas giant. The location of many asteroids together in the main belt is a useful result for scientists who are wanting to study the early phase of planet building. They provide a way of sampling the rocky planets at the time they were forming, as they preserve information that the planets no longer recall. It is tricky to use the large, rocky planets to study this early era of Solar System history because they continued to evolve and chemically change for a long time after they formed. In fact, they continue to do so at the present day. The planets became big enough to experience large-scale geological processes, create atmospheres and undergo complex internal processes that acted to change their interiors and exteriors. The geological evolution of many of the asteroids, on the other hand, was all but halted just after they came into being, preserving evidence of the exact time of planet formation and giving scientists access to this key stage of Solar System growth.

The difference between comets and asteroids

Presented here is what can be considered as the 'classic' view of the formation of comets and asteroids. Their differences – fluffy, icy and fragile comets versus hard, dry and solid asteroids – mark a fundamental distinction between the vast majority of these objects. Their structural and chemical contrasts reflect the place and the conditions under which they formed – cold, outer disc versus hot, inner disc. This classic view has been shaped over many years and, for the most part, it is broadly accurate. However, in reality, and particularly in light of recent space mission data, the picture is turning out to not be quite so simple. Recent findings have shown that there can be some overlap in the composition and structure of comets and asteroids. Put simply, observations have been made of icy asteroids and rocky comets, findings that have tested the classic view of these objects. How can an asteroid contain ice when it formed in the hot inner Solar System? This is an area where scientists are breaking new ground and learning that these weird and wonderful, varied and unexpected objects have new secrets to reveal about an important time in Solar System history. Here we will unravel more about the long yet fascinating comet and asteroid stories. What can be said for now is that the Solar System's early days are not as straightforward as the classic models allow, but this certainly makes for a more interesting read.

The comets

The majority of comets formed beyond the so-called 'snow line' of the solar nebula – a region where it is thought to have been cold enough for volatile species such as water, ammonia, methane, carbon dioxide and carbon monoxide to condense as ice. We will use the term 'AU' as a unit of measurement in this book. It stands for 'astronomical unit' and is equal to the distance between the Earth and the Sun. AUs are a handy way to discuss space distances, since 1AU is around 149.6 million km (93 million miles). During the formation of the planets the snow line is thought to have been located at around 2–3AU, but today it is located around 5AU because of the changing conditions of the Sun and solar nebula over time.

The name 'comet' derives from the Greek word for 'long-haired', signifying the long tails comets produce as they pass near the Sun. Such tails are said to become 'active' as the comet crosses the snow line to enter the inner Solar System, causing it to release gas and dust as its nucleus is heated. Comets can produce magnificent tails, which can extend hundreds of kilometres from the nucleus in some cases, sometimes visible from Earth with the naked eye. The coma – an atmosphere around the comet produced by an envelope of gases that have sublimated from the nucleus as it is heated – is also a key feature that gives the comet a blurry outline when viewed by telescope.

While the dirty snowball analogy is useful, as we discussed, another good way to think about comets is as the fluff of the Solar System: this helps to describe their fragile texture, something that is directly inherited from the environment in which they formed. When comets are described as a scoop of the early solar nebula it's because they collected up a mixture of the fragile dust, primitive organic compounds and ices that it contained. These are said to be 'primitive' components as they are the oldest ingredients in the Solar System, representing the material that formed the early solar nebula. However, the true picture of comets is probably much more complex. They are almost certainly formed from a mix of different materials that were gathered together from a wide range of locations around the early solar nebula. The early-formed swirling solar nebula-mixed Solar System components formed in different locations and times – at diverse temperatures and pressures – across large regions of space. Although it can be said that the majority of comets formed in the 'outer disc', in reality this is a very large region to consider and it is therefore unlikely that all the comets formed in precisely the same way. Comets are valuable scientifically for exactly this reason: not only do they contain a sample of the main ingredients required to make a Solar System – the earliest gas and dust – they also sample a large region of the early solar nebula.

One of the most important things about the comets is that their early-formed, fragile ingredients have never been heated, because the vast majority of comets have remained in the Solar System deep freeze far from the Sun since they formed. Of course, some occasionally exit their cometary home to make a journey close to the Sun, and every time a comet passes close to the Sun it is altered slightly because its volatiles are heated and sublimated, carrying away some rock dust. Sometimes the results of this heating occur explosively, with strong jets of dust streaming from the comet. Not only do these processes dry out the comet surface, removing volatile materials, but they also take away a great deal of the comet mass. A comet can lose a layer of material a metre (3ft) thick from its surface with each passage close to the Sun. So, the fewer times a comet has approached the Sun, the better for studying their early-trapped ingredients. This means that the comets remaining far away from the Sun, or those that enter the inner Solar System for the first time, preserve their ancient solar nebula ingredients the best, making them extremely valuable for our understanding of where everything in our Solar System originated.

Comets currently reside in two main locations, both of which happen to be very far from the Sun in the cold, far outer reaches of the Solar System. One of the cometary neighbourhoods, the so-called Oort Cloud, is so far out that it is barely within the gravitational grasp of our star. This is the region of the Solar System that we likened to the rural communities existing in the greenbelt surrounding a city, each one located a great distance from the others, but all looking similar in nature. The closer comet neighbourhood is the Kuiper Belt, sometimes known as the Edgeworth-Kuiper Belt, which is still 10 times further from the Sun than the asteroid belt. The Kuiper Belt acts much like the outer suburbs of the city, not quite the very edge, but a journey into the city would be a long and arduous one.

The Kuiper Belt was first observed in 1992, with its existence having been predicted in 1951 by the Dutch-American astronomer Gerard Kuiper. It was found to be made up of hundreds of millions of icy objects which inhabit a region 30 to 55AU from the Sun. The Kuiper Belt includes not only comets but some icy dwarf planets such as Pluto, highlighting the fact that it is made up of a range of different-sized objects.

The Kuiper Belt comets are on fairly stable and generally elliptical orbits. However, they can be affected by the gravity of the planets, which can disrupt them from their stable orbit, deflecting them into the inner Solar System where they pass close to the inner rocky planets, or instead end their existence in a fiery solar collision. The comets that exit the Kuiper Belt are known as the short-period comets, with orbital periods of less than 200 years; that is, they take less than 200 years to complete one orbit of the Sun. We can think of these as the part-time commuters who spend most of their working week at home, only occasionally making their laborious journey into the city. Short-period comets orbit the Sun along the ecliptic, the same as the planets, and are responsible for the well-known comets such as Halley.

At around 50–60AU, there exists a subclass of the Kuiper Belt known as the scattered disc. This is thought to be the source of the Jupiter-family comets, so-called because their current orbits are primarily influenced by Jupiter and have orbital periods of less than 20 years. Comets 67P/Churyumov–Gerasimenko (hereafter 67P/C-G), visited by the *Rosetta* mission, and 81P/Wild2, visited by the *Stardust* mission, are Jupiter-family comets that are both just a couple of kilometres in diameter. However, there are some fairly large objects in the scattered disc, including Eris, which reaches 2,300km (1,429 miles) in diameter and whose large size and presence at this huge distance from the Sun in part caused the reclassification of Pluto as a dwarf planet.

The farthest comet reservoir, the Oort Cloud, is composed of two regions, both of which lie in interstellar space beyond the heliosphere – that is, outside the solar magnetic field. One of these is the disc-shaped inner Oort Cloud, otherwise known as the Hills Cloud, and the other is an outer spherical cloud of comets. Interestingly, scientists have not yet confirmed the Oort Cloud's existence because the icy objects that it is thought to contain are small, and so far away that they can't be seen. The Oort Cloud's presence was predicted in 1950 by the Dutch astronomer Jan Oort in order to explain some infrequent icy visitors to the inner Solar System. These are the long-period comets, so-called because of their long orbital periods, which reach thousands to millions of years. The interesting thing about these comets is that they enter the inner Solar System on very elliptical orbits from different and random angles. The only way that the presence of such objects can be explained is if there is a vast, distant reservoir of comets further out than the Kuiper Belt that surrounds the entire Solar System, acting as a sort of cloud. Hence, the Oort Cloud is assumed to be there.

At its closest, the Oort Cloud is thought to orbit at 1,000AU but with a radius of 50,000 to possibly 200,000AU, containing billions to trillions of comets. Some estimates suggest that the outer edges of the Oort Cloud may even be closer to the next star system than to our own Sun. These icy objects really are those remote rural farms that are as close to our Solar System city as they are to the neighbouring city, yet still a great distance from both. It is the gravitational influence of passing stars, and even the Milky Way itself, that can perturb and tug at the outer Oort Cloud comets, loosely bound to the Sun because of its weak gravitational influence at this distance, disturbing them from their orbit and sending them careering into the inner Solar System. Perhaps if we stretch our Solar System–city metaphor a bit further we can suggest that these comets are like the farm workers who seldom get a night off from their hard labouring, but when they do they travel a long way to the city for a wild night out. Some famous long-period comets that have taken a similar journey include the beautiful Hale-Bopp, discovered in 1995 and one of the most widely observed and brightest comets in history, and Comet ISON, which dramatically exploded when it transited too close to the Sun in 2011.

The Oort Cloud comets are thought to have formed close to where the orbits of Jupiter, Saturn, Uranus and Neptune are today, whereas the Kuiper Belt comets were initially formed from a disc-like distribution of icy objects beyond the orbit of Neptune and Uranus. You might have noticed from this that the relative positions of the Oort Cloud and Kuiper Belt comets are seemingly the 'wrong' way around compared with their current locations. Well, in the early Solar System very soon after the planets had formed, the gas and ice giants are not thought to have been in their current positions. Instead they were in a more compact configuration, arranged much closer to one another than they are now. Uranus is even thought to have been on a larger orbit than Neptune, *i.e.* they were reversed compared with their current positions. However, fairly early on in the Solar System's history, a few hundred million years after it formed (in its 'teenage years'), an event occurred that completely rearranged the positions of many of these objects, including the giant planets, the comets and the asteroids. We'll come back to this later, but first we need to visit the asteroids.

The asteroids

Many of the asteroids are thought to have formed roughly where they are located at present, in the asteroid main belt, much like the inner rocky planets of Mercury, Venus, Earth and Mars. As such, these asteroids contain many of the same main Solar System ingredients as the inner rocky planets – rock minerals and metals formed at high temperature next to the Sun, very little water and only small amounts of organic matter. We can think of these objects as 'classic' asteroids, *i.e.* they contain what we expect based on classic models of Solar System formation as described above.

Most asteroids currently reside in the main asteroid belt, where there are thought to be up to 2 million of them over 1km (0.6 miles) in diameter and millions more smaller ones. Nevertheless, if you added all of them together, their mass would still be less than that of the Earth's Moon. The asteroids in the main belt range from pebble size to much bigger with the largest being Ceres, at 950km (590 miles) in diameter. It is so large that it has been reclassified as a dwarf planet, the smallest and closest dwarf planet to Earth. After Ceres, the next largest object is asteroid 4 Vesta, which is over 500km (310 miles) in diameter. The hefty Ceres and Vesta were the focus of the NASA *Dawn* mission that launched in 2007.

Interestingly, as recently as the 1990s, scientists had discovered only about 10,000 asteroids, but by the beginning of the new millennium that number had doubled to 20,000. At the present day, over 700,000 asteroids with known orbits have been discovered. For about 100,000 of these, scientists have fairly detailed information about them, including roughly what they're made from. Using telescopes, it is possible to estimate the composition of the surface of an asteroid without having to leave the confines of Earth. For example, the size of the rock particles forming the asteroid, and their mineralogy, can be found by taking measurements at different wavelengths of light, from ultraviolet to infrared. Although such methods don't allow for the entire surface of an asteroid to be mapped in detail, as they lack the spatial resolution, they still give scientists a good idea of the overall composition of the different objects that are out there, allowing them to be placed into groups. This is how scientists know that there are some rocky asteroids, made up of silicate minerals, some metallic asteroids and some that are a mixture of both silicates and metals.

It's not really correct to consider the entire asteroid belt as a single entity because the composition of the asteroids – at least the largest ones, over 100km (62 miles) in diameter, which are the easiest to measure from Earth – vary depending on where they are located. It turns out that there are clear compositional and structural differences between the asteroids found in the inner asteroid belt and those found in the outer. The differences observed across the belt are almost certainly reflective of where the individual asteroids formed. Some of the asteroids are even thought to have formed in a far-off location in the solar nebula before finding their way into the belt, but more on that shortly. Understanding where individual asteroids formed involves having knowledge of what their different compositions reflect, and what this might say about where they came from.

First, let's take a little step back to understand what the overall composition of the asteroids – silicate rock versus metal – can tell us about their history. The silicate rock minerals found in asteroids are very common on Earth and other planets, too; they constitute over 90 per cent of the Earth's crust. Silicate minerals form when two basic elements, silicon and oxygen, are combined. Different silicates can incorporate into their structure a whole range of other elements: iron, magnesium and aluminium, to name but a few. These elemental ingredients were all available in gas form in the early solar nebula. The silicate minerals and metals formed by condensation of these elements as the cloud cooled down soon after the birth of the Sun. The order in which the minerals condensed is predictable and can be calculated using geochemical equations. Because metals have the highest melting point, they are the first to condense when the gas cloud starts to cool, followed by silicate minerals with progressively lower and lower melting points.

With this information in mind, the different classes of asteroids within the main belt can be defined based on their colour and reflectivity as viewed by telescope. These features allow scientists to estimate how much metal, silicate, carbon and volatiles individual asteroids contain. Such observations reveal that the inner half of the asteroid belt, nearer the Sun, contains objects with high reflectivity. These asteroids are also almost completely devoid of water and other volatiles, tending to be rockier and more consolidated – what we would consider a 'classic' asteroid. They are defined as either V-type, after asteroid 4 Vesta, or S-type, *i.e.* siliceous (containing silicates) or stony. Their compositions reflect their formation in a high-temperature environment, almost certainly quite close to the Sun, which explains why they don't contain volatiles. The very inner edge of the asteroid belt contains the M-type asteroids. These are highly reflective and are found to contain a very large proportion of metals, making them extremely dense.

The asteroids of the outer belt are dark in colour. Some of them are the C-type asteroids that are rich in water, but the water is in the form of ice because they are still a long way from the Sun, nearly at the orbit of Jupiter. These asteroids also contain carbonaceous organic matter and fragile dust. Then there are also the D-, P-, B- and G-type asteroids that seem to be closely related to the C-types; they contain the same main ingredients in different proportions. Yet, these are the very same ingredients contained within the comets: fragile dust, organic matter and volatiles. And the volatiles are present as ices! This is not a feature we should be expecting in asteroids according to the classic model. They should have been formed in a region of the disc that was too hot to preserve ice.

It is these icy, carbonaceous asteroids that lead scientists to question whether they can really be called asteroids, as their compositions certainly don't fit the classic model. Apart from the fact that they are located in the asteroid belt, they look like comets. If one of them were to journey closer to the Sun then it would almost certainly become active and produce a tail, just like a comet, because of all the icy volatiles it contained. In fact, scientists think that these asteroids may have originally formed far out on the cold edge of the early disc, just like the comets. They were, however, subsequently diverted, or gravitationally thrown, into the inner Solar System shortly after their formation. Instead of scooting close to the Sun, or ending up colliding with it, they were caught up in the edge of the asteroid belt en route. We'll come back to these fascinating objects a few times throughout this book because they are an important focus for scientists who study the early Solar System. After all, they contain the same primitive ingredients as the comets, but they are much closer to Earth, so easier to get to.

In between the inner and outer asteroid belt there are many more C-type asteroids, but they are joined by B-type asteroids. Bennu, a B-type asteroid, will be visited by the NASA *OSIRIS-REx* sample return mission that launched in 2016. Asteroids like Bennu are intermediate in composition between those of the inner and outer belt and are likely to have a mixed origin depending on their individual composition.

The dance of Jupiter – accounting for the asteroid belt

Scientists have proposed a model to account for the observed general architecture of the asteroid belt and it is known as the 'Grand-Tack Model'. In this model, early on in Solar System history, the newly formed Jupiter is believed to have shifted around, migrating inward towards the Sun and back out again. The 'tack' part of the hypothesis name is because its movements can be likened to a sailboat 'tacking' against the wind. Because Jupiter is so large, its 'do-si-do' dance is thought to have torn apart an infant asteroid belt, scattering objects all over the Solar System, including completely out of it and into the Sun, before repopulating it again. The newly formed asteroid belt contained inner asteroids that journeyed from closer to the Sun – explaining their rocky and volatile-free composition – and outer asteroids that were dragged in from beyond the giant planets, explaining their more 'primitive' icy compositions.

The complex architecture of the asteroid belt shows why it's important scientists study many different types of asteroids. Studying one asteroid can't necessarily inform them about the population as a whole, neither can studying only one type. Different classes of asteroid preserve information about varying locations and times in the history of our Solar System. Asteroids, therefore, can inform scientists about the state of the Solar System not just during the formative stages of planet building, but even earlier in time.

The Nice Model – accounting for the Oort Cloud comets, and the surface of the Moon

You'll recall that we still need to explain how the Oort Cloud comets ended up being so far from the Sun when they apparently formed closer to it than the comets of the Kuiper Belt. The best way for scientists to account for this observation is if the Solar System experienced a violent reconfiguration that they think is thanks, once again, to Jupiter. The model scientists proposed was originally presented as a trio of papers published in the scientific journal *Nature* in the early 2000s and is now called the 'Nice Model', after the city in France where the astronomers were working. However, apart from just trying to account for the configuration of the comets, scientists had been looking for a way to explain some other key observable Solar System features. The most obvious of these was the pockmarked appearance of the Moon, which can be clearly seen from Earth, even with binoculars. The visible scarring of the lunar surface is caused by a vast amount of overlapping impact craters that disfigure it. Scientists first needed to know how old these impact craters were so that they could figure out if the craters were the result of impacts that occurred over the entire history of the Moon or just in a short period of time.

Using laboratory measurements of rock samples returned by the *Apollo* missions from some of these craters, it was found that the ages of the craters fall between 4.1 and 3.8 billion years ago. At this time, our Solar System should have been going through a relatively quiet phase, around 500 million years or so after it formed. The rather chaotic planet-building phase was over and the rocky inner planets had, more or less, settled into their current orbits. The age and density of the large craters scarring the lunar surface is clear evidence for a step away from this quiescent time, an era that resulted in the Moon's surface being pummelled by large, high-speed impacts of objects from space. This period in our Solar System's history is called the Lunar Cataclysm or the Late Heavy Bombardment, and although it was recognised most easily on the Moon, because the craters are well-preserved there, it affected the entire inner Solar System.

The fact that we don't see evidence of a similar bombardment on Earth doesn't mean our planet escaped violent collisions. It is certain that Earth was also pummelled by incoming space rocks around 4 billion years ago, but not much evidence for this bombardment exists on our planet at the present day. The reason for this is that Earth has continually changed its appearance, on geological timescales anyway, because of plate tectonics and other geological processes like weathering and erosion. Some of this geological activity is even responsible for the Earth's surface being melted, destroyed and re-made many times over.

The outermost shell of the Earth can be likened to a massive jigsaw, but with seven major pieces, and some smaller

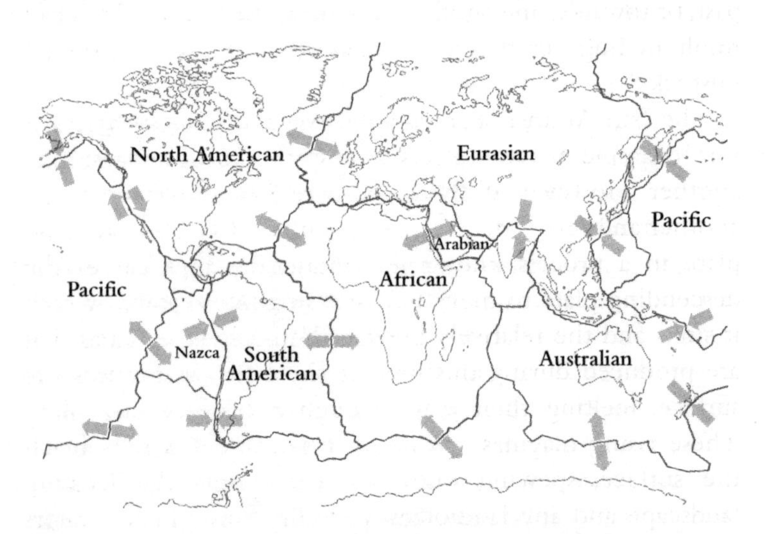

***Figure 3** The tectonic plates of Earth. A diagram to show the positions of the main tectonic plate boundaries, with arrows indicating their relative directions of motion in relation to each other.*

but no less significant pieces. The Earth's major jigsaw pieces, encompassing areas greater than 20 million km^2 (7.7 million sq miles), are the tectonic plates that carry the landmasses. But unlike a normal jigsaw, these plates jostle for position as they attempt to seamlessly float on the Earth's somewhat squidgy underlying mantle. The mantle is solid rock, but because of the extremely high temperature and pressure conditions it is subjected to, it flows over long geological timescales (thousands to millions of years), which causes the jigsaw pieces lying above to move sluggishly along with it. Still, the dance of the plates on top of the mantle can hardly be described as graceful, with most of the action happening at the boundaries between the plates. The activity that occurs at these boundaries is dependent on the relative density of the plates and their directions of motion. The density of a plate depends on what type of rock it is composed of, with rocks made on the ocean floors being much denser than the rocks that compose the continents and big mountain chains such as the Himalayas. This is, in part, what makes the oceans deep and the mountains high. When plates attempt to move past, or towards, one another they often get stuck, which can result in huge earthquakes when they eventually become unstuck.

The San Andreas Fault in the western United States is one example of two plates attempting to glide past one another. However, if two plates are headed towards one another and one is denser, then it sinks below the less dense plate in a process known as subduction. This causes the descending plate to melt the Earth's mantle through which it sinks and the relatively buoyant liquid rock magmas that are produced during this process rise back up towards the surface, melting their way through the overriding plate. These rising magmas can produce volcanoes if they reach the surface, spewing out lava that covers the existing landscape and any landforms, including any impact craters that might have been present.

Over time, plate tectonic processes act to re-surface the globe such that the evidence for asteroid and comet impacts is lost to the geological record, either eaten up in the Earth's interior or covered over by new rocks. This is why the evidence for the Late Heavy Bombardment has been all but lost on Earth, whereas, due to its lack of plate tectonics, the Moon has clearly preserved the entire violent history. Thank goodness for that, otherwise scientists might have had a lot more trouble piecing together an important phase of planetary evolution.

Once scientists worked out that the Solar System experienced a major phase of carnage after the planets had formed, they needed to find a way to account for this violent time. The Nice Model, despite not being able to explain all of the intricacies of our Solar System conditions at this time, does present a hypothesis that can help explain, at least partly, why this chaotic period of Solar System history might have occurred.

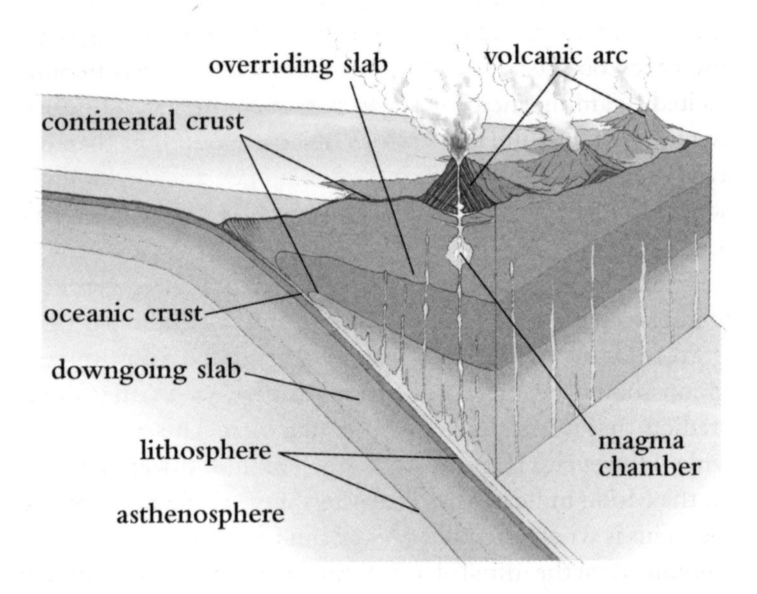

Figure 4 *Plate tectonics in process. A cross-section through the lithosphere (the rigid outer portion of the Earth hosting the tectonic plates) and asthenosphere (the underlying easily deformed layer which the tectonic plates slide over) demonstrating how tectonic plates undergo subduction – where one plate sinks beneath another – to create volcanic arcs on the overriding plate.*

At around 700 million years after Solar System formation, computer models that simulate orbital mechanics during this time show that Jupiter and Saturn briefly experienced an orbital resonance as Jupiter was slowly migrating inwards to the Sun. An orbital resonance is where two orbiting bodies exert a gravitational influence on each other because their orbital periods (the time it takes them to orbit the Sun) are related by a ratio of two small integers, in this case 2:1 (*i.e.* Saturn orbited the Sun once for every two Jupiter orbits). Effectively, these large planets hit a gravitational sweet spot that increased their orbital eccentricities, causing their orbits to become less circular. Because of the huge size of these planets compared with the objects surrounding them, Jupiter and Saturn acted to rapidly destabilise the entire Solar System. Jupiter forced Saturn outwards away from the Sun, and its movement caused the orbits of the ice giants, Uranus and Neptune, to become unstable. This sent them ploughing into the surrounding outer disc of icy objects. The result was that the icy objects became excited – gaining energy – and were scattered all over the Solar System. Not only did this send icy objects out even further, but it also sent them in the opposite direction, towards the inner Solar System. The total mass of icy objects calculated to have been scattered into the inner Solar System at this time is consistent with estimates for the number and size distribution of impacts seen on the Moon's surface.

From this description, it seems that the craters on the Moon should be the result of comet impacts, as the model predicts that it was the icy objects that were scattered far and wide. However, chemical studies on some of the large basins on the Moon indicate that the craters were made by asteroids, too. This is where the Nice Model can help again as, although it predicts that the initial phase of the Late Heavy Bombardment was probably dominated by cometary impacts, it indicates that the final objects to be flung around the Solar System were probably the asteroids. Overall, the model predicts that the mass of asteroids and comets hurtling into the inner Solar System may have been roughly equal over the course of the Late Heavy Bombardment, which may have lasted anywhere between 10 and 150 million years. However, it is a number that is hard to determine as the original mass of the asteroid belt is currently unknown.

Asteroid or comet? It's anyone's guess

A tumultuous reorganisation of the Solar System, as described here, hints at the complication of trying to place the formation locations of comets and asteroids in neat boxes. Even if objects were formed in the far outer reaches of the Solar System, they haven't necessarily ended up staying there permanently, and vice versa. Much like the inhabitants of our Solar System-city having moved around throughout their lifetimes, perhaps starting their young life in the suburbs, migrating for work to the bright lights of the city and eventually back out again as they grow older, the Solar System objects have, in some cases, experienced a similar life. The movements of objects around the Solar System early on means that care needs to be taken in assuming too much about the composition and history of an object based on its current location.

Over recent decades it has become apparent that the classic models for the formation and composition of comets and asteroids need revision. The models are not completely inaccurate – they can account for the broad history of many of the small objects out there but perhaps nature is less concerned about the lines scientists have drawn, with some comets and asteroids not adhering neatly to the rules. Although the two boxes that comets and asteroids have seemingly been placed in are convenient, they are hiding a more interesting Solar System story. Certainly, one box contains primitive objects, which include the majority of comets but also some asteroids that have high ice and volatile abundances, along with high carbon contents but little metal. The other box contains more evolved objects, represented by many of the asteroids, that are rocky, have little or no ice or volatiles and low carbon contents. However, observation has shown that there are small space objects that don't fit into these boxes, being neither 100 per cent asteroid nor 100 per cent comet. It is certainly a possibility that asteroids and comets display a continuum of compositions, extending from primitive, classic 'comet-like' to more evolved, rockier, classic 'asteroid-like' and everything in between. In fact, some asteroids have been observed that display 'cometary' behaviours; they are 'active', emitting sporadic streams of dust as they orbit the Sun. When such objects are observed in the asteroid main belt they are called 'main belt comets', but it is not clear if their activity results from sublimation of ice, as in comets. Because of this, they might not be *true* comets, so some scientists simply refer to them as 'active asteroids'.

At the other end of the scale there are the bizarre 'asteroid-like' comets that are notable for their lack of cometary behaviour. Comet C/2013 P2 and C/2014 S3 were observed travelling in towards the Sun, where temperatures are obviously high enough to produce cometary activity. But these comets showed no dust tail at all, and C/2014 S3 even had a colour observed by telescope that more closely resembled an asteroid. The interpretation is that these comets might have started their life close to the Sun, like the asteroids, and been flung way out of the inner Solar System to join the 'true' comets that were formed in the outer disc. Thinking back to the Nice Model, this is hardly unexpected, once we have understood the movements of the large planets and the havoc they wrought throughout the inner Solar System. The question still remains: should these objects be classed as asteroids or comets? It's anyone's guess. They may be located on a cometary orbit, but they seem to resemble 'classic' asteroids in appearance and composition. These objects are probably not alone either: it is almost certain there are others out there that share a similar history.

As scientists continue to observe and sample other small Solar System objects, they are likely to find more unexpected results just like these. To make matters even worse, a comet that has passed close to the Sun many times will gradually lose its volatile material, which acts to dry out the comet and turn it into a chunk of rock, thus essentially transforming it into an asteroid. The neat boxes suggested for these objects really don't provide a good fit for so many small Solar System bodies but it doesn't mean that they aren't a useful starting point. Clearly, however, if the early Solar System is to be understood and mapped out in detail, many more of these objects will need to be studied more closely.

Materials older than the Sun

This is probably a good point to mention something I've kept somewhat hidden up to this point. I've said that comets and asteroids are made up of the earliest materials that formed *within* our Solar System. While this is true, they also contain some material that is *older* than our Solar System – what is referred to as 'presolar' material. From the name, 'pre' and 'solar', you can probably guess that presolar materials come from a time before the Sun. When our early nebula formed it swept up tiny grains that had previously been ejected from other stars that came before – exotic stardust. These grains are made of particularly hardy substances such as diamond, graphite, silicon carbide, titanium oxide, spinel and corundum. Presolar grains made of these substances managed to survive the violent birth of the Sun such that they became part of the structure of the earliest objects that formed out of the swirling cloud of gas and dust – the comets and asteroids. This means that not only can these special Solar System objects allow scientists to probe back to the beginning of the Solar System, but they can also allow them to voyage to other stars in the galaxy.

Comets and asteroids near to the Earth

Comets and asteroids can be observed with telescopes, which can often tell scientists a great deal about their composition, shape, orbit and behaviour as they make their way around the Sun. If scientists want to visit and obtain a rock sample from an asteroid or a comet, it proves to be a little bit trickier. Luckily, it isn't necessary to go all the way to the asteroid belt, or even worse, out to the far outer reaches of the Solar System to the Kuiper Belt or Oort Cloud, to visit these objects. Rather usefully, the Solar System has a way of diverting asteroids and comets from their normal orbits on to one that brings them closer to us. As a collective, these diverted objects are called near-Earth objects, or NEOs, but they can include near-Earth asteroids (NEAs) and near-Earth comets (NECs). By definition, such objects must have orbits within 1.3AU. It is estimated that only about one per cent of NEOs are comets, suggesting that the population is dominated by asteroids. However, around 15 per cent of the NEA population are ex-comets, or dormant comets – those that have lost their volatiles and have become, simply, a lump of rock from repeated transits near the Sun.

NEOs are slightly easier to approach with a spacecraft than the same type of objects in the asteroid belt, Kuiper Belt or Oort Cloud. Nevertheless, NEOs can pose a big problem as their altered orbits around the Sun can place them on one that intersects that of Earth. A large asteroid or comet colliding with Earth would almost certainly threaten life across the planet, independent of where it impacted. The risk of an impact in the future is taken so seriously that there is a monitoring system in place called Spaceguard, itself a loose term for a number of efforts around the globe concentrating on tracking potentially hazardous space objects. Spaceguard, named after the fictional early-warning system in Arthur C. Clarke's book *Rendezvous with Rama,* began in earnest in 1998 to study NEOs that pose a threat to Earth. Using telescopes, and there's even a spacecraft-based telescope tracking NEOs called NEOWISE (Near-Earth Objects Wide-field Infrared Survey Explorer), these efforts track and monitor NEOs, particularly objects over about 140m (460ft) in diameter. It is estimated that this size, or larger, has the potential to cause devastating effects for life on Earth on a global scale in the event of an impact. Over 15,000 NEOs have been found, of which around 1,800 appear to be on a potentially hazardous orbit, defined as an orbit that comes especially close to Earth: within 19.5 lunar distances, or 7.5 million km (4.6 million miles). This might make the situation sound a little worse than it is, though, as scientists still don't have a great deal of information about some of these objects and many of their orbits are not yet perfectly defined. As more data are collected, in particular data that allow for a more accurate calculation of the orbit of these objects, then NEOs could be removed from the list if it transpires that they will pass the Earth with enough distance that a collision is impossible. NEOs can spend as long as a few million years in a near-Earth orbit before eventually being ejected from the Solar System altogether, colliding with the Sun or, less commonly, colliding with a planet.

The possibility remains that an NEO could be found to be on an exact course for Earth in the near or distant future on human timescales. While large asteroid and comet impacts are low-probability events, with objects 1km (0.625 miles) in diameter hitting the Earth on average once every 500,000 years, smaller objects hit us more frequently and can still cause considerable problems. At the present time, there are no firm plans in place to counter the possible collision of a large space object with Earth. At best, the region that is predicted to take the brunt of the impact might be evacuated. However, we now know what happened to the dinosaurs when a large space rock hit the Earth 65 million years ago; it's not the impact itself which is necessarily the problem, but the long-lasting environmental after effects. How to mitigate such a disaster is, you'll be glad to learn, a key active area of research in the NEO community, even if some of the mitigation plans may at first sound a bit like science fiction; think *Armageddon* and similar movie scenarios. We'll learn more about these in Chapter 10. What is important at this time is that scientists are working on understanding more about the small, but potentially hazardous, space rocks that are out there. They must understand what they might be dealing with, what the object is made of and how fragile or solid it is. An icy dust comet heading our way might not pose as great a risk as a huge lump of metal, for example.

Apart from the obvious scientific value of NEOs, these space objects may also serve a purpose economically, as we'll see in more detail in Chapter 9. Among all the important early Solar System ingredients they contain, which allow us to learn about the formation of the Solar System, asteroids and comets also contain an abundance of valuable resources, including platinum, gold, iron, nickel, rare earth metals and water. The great thing about asteroids, in particular, is that they often contain these things in abundance. In fact, it's possible that the platinum contained in an asteroid just 1km (0.625 miles) in diameter could be worth $150 billion. It might be economical to mine asteroids for metals if this cost is less than trying to mine the same deposits from dispersed resources on Earth. These small and varied space objects really do have a lot to offer us, as long as they don't wipe us out before we can utilise them.

As we've seen, comets and asteroids have experienced a long and dramatic history in our Solar System, and they might have played a key role in the history of our own planet by delivering the ingredients for life on Earth, of which there is more discussion in Chapter 5. At the same time, comets and asteroids might also be important in our future if we can successfully exploit them for their resources for use on Earth, or use them to allow us to travel further into the Solar System. However, the only way asteroids and comets can play an important role in our future, and one in which they aren't responsible for our demise, is if we study them in detail to understand how they were made, what they contain and how they will behave on their long orbits around the Sun. Only then can we make sure that these space rocks remain things of beauty to be held in awe, rather than harbingers of doom.

CHAPTER 3

Comets and Asteroids on Earth

Over the course of the past 50 years, space instrumentation has become more and more advanced as humans have pursued a varied number of different objects in our Solar System to image, measure and sample. Humans have successfully placed a fully functioning rover on the planet Mars to roam over its surface, drilling and collecting samples to analyse on-board its cargo of scientific instruments. A sophisticated scientific laboratory has also been sent into space on a decade-long journey to catch up with and land on a speeding comet to perform analyses of its rocks, ices and gases. And this is to name just a few of the more recent highlights of space exploration. However, despite these advances and amazing achievements, the best and most easily controlled scientific instruments exist on Earth. The problem is that these Earth instruments can't be sent into space very easily – they are too heavy and sensitive to launch aboard a rocket and they need near-perfect conditions to perform with precision and accuracy. The space environment is not a friendly place, with substantial extremes in temperature and pressure, conditions that are not suited to delicate and, sometimes, temperamental laboratory instruments.

The result is that there are often many advantages to bringing space rock samples back to Earth for careful, considered and precise analysis, as opposed to attempting to launch advanced laboratory instruments into space. The major problem, however, is that collecting rocks in space and bringing them safely back to Earth is no simple task either. In fact, sample return from space has only been achieved a few times: from the Moon with the *Apollo* and *Luna* missions in the 1970s, from asteroid Itokawa with the *Hayabusa* mission and from comet 81P/Wild2 with the *Stardust* mission. Although hundreds of kilograms of Moon rock have been returned to

Earth, the *Hayabusa* and *Stardust* missions only returned minute amounts of rock sample – dust-sized fragments to be precise. Still, tiny samples are certainly better than no samples, as even small rocks can hold an immense amount of information in their structures – secrets that scientists can unlock with their highly specialised scientific instruments on Earth.

The *Stardust* mission, in particular, achieved a great deal in furthering our knowledge of the composition of comets. The comet dust samples it returned to Earth will keep scientists busy for many decades to come, despite their limited mass. We will learn more about this mission, and the precious samples it collected, in Chapter 7. Luckily, there are future plans for collecting rocks from space, with some missions already on their way and others awaiting funding. These missions include visits to asteroids, the Moon and Mars, and while they may all be risky endeavours with no guarantee that they will achieve their goals, it is good to know there is hope for the return of samples from space for Earth-based analysis in the future.

The arrival of space rocks on Earth

Luckily, it turns out that there's another way to obtain samples of space rocks and it doesn't even involve leaving the safe confines of Earth. This is because space rocks naturally fall to Earth as meteorites all the time. In fact, some 40,000 to 80,000 tonnes of space rocks fall onto our planet each year. These free space samples can be likened to cosmic Kinder Eggs – they are packed with celestial prizes, information about our Solar System. Meteorites can include samples of asteroids, comets and other planets, most of which haven't been sampled by spacecraft yet.

Of the thousands of tonnes of space rock arriving on Earth each year, the majority are quite small, mostly dust-sized, of which we'll learn more in Chapter 4, but some individual rocks can be quite large. Some of the largest stony meteorites to arrive on Earth have been up to 60 tonnes in weight, which is about the same as five double-decker buses. Meteorites can originate from anywhere in space, but it tends to be rocks from asteroids that are most commonly found on Earth as pebble-sized pieces, although pieces of comets and planets can also appear. Chunks of asteroids can end up hurtling towards Earth after they have broken off from their larger parent asteroid in space, often during collisions with other space objects, which can cause them to break apart completely or for small pieces to be knocked from their surfaces. In space, once these small samples of asteroids have broken away from their parent rock they are called meteroids and they can spend hundreds, thousands, perhaps even millions of years travelling through space until eventually colliding with a moon, a planet or the Sun. As the rock enters the atmosphere of another planet it becomes a meteor and if and when these pieces reach the Earth's surface, or the surface of another planet or Moon, they become meteorites. There is nothing magical about an incoming space rock turning into a meteorite, it is simply a name the rock receives when it becomes stationary at the surface of the body it meets.

If all these space rocks naturally arrive on Earth for free, then you might wonder why scientists bother going to the trouble of visiting space to attempt sampling at all. Despite the fact that the rocks falling to Earth sample a much wider range of Solar System objects than humans can visit in many lifetimes, these samples tend to be biased towards those that can best survive the harsh effects of atmospheric entry. The issue arises because of the extreme temperature and pressure changes experienced by a rock, or any object, during atmospheric entry from space to Earth, variations that are large enough to totally obliterate a rock in many cases.

Temperature changes during atmospheric entry occur as a direct result of the high incoming velocity of the object, which can be anywhere from around 10km/s to 70km/s (25,000mph to 150,000mph). The problem for the incoming space rock when travelling at these hypersonic velocities is that the atmosphere can't move out of its way quickly enough. Such an effect is absent as a rock travels through space, simply because space is a vacuum so there are too few molecules present to knock into each other. A rock travelling through an atmosphere has a buffeting and compressing effect on the molecules it encounters, causing them to pile up and dissociate into their component atoms. These atoms ionise to produce a shroud of incandescent plasma that is heated to extremely high temperatures – up to 20,000°C (36,032°F) – and envelops the space rock, causing it to become super-heated. The result is that the rock appears to burn and glow in the atmosphere; what we might call a fireball or a shooting star, depending on its size.

The effects of this process bring about a notable physical change to the incoming rock, one that actually makes it easier for us to identify when it becomes a meteorite on the surface of the Earth. That is, the formation of a fusion crust, which develops as the rock penetrates the lower atmosphere and is slowed down and heated by friction with the air. The outer portion of the rock starts to melt and the mixture of liquid and gas that forms is swept off the back of the meteorite, taking the heat with it. While this process is continuous and means that the heat cannot penetrate the rock (thus acting like a heat shield), when the temperature finally drops, the molten 'heat shield' solidifies as the last remaining liquid cools at the rock's surface to form the fusion crust. The resulting dark, often shiny, rind on meteorites is a distinctive feature that can often be used to help identify them and to tell them apart from terrestrial rocks. The formation of the fusion crust protects the internal parts of the meteorite from the worst effects of the heat, preserving the composition of the parent asteroid, comet or planet from which it originated. However, although meteorites closely resemble their parents, they are not an exact match. In the process of forming the fusion crust, the rock loses some of its more volatile components as they are boiled off with the extreme changes in temperature experienced in the outer layers of the rock. The only way to obtain a 'perfect' sample would be to collect one directly from a space object and return it in a spacecraft. However, since meteorites are free samples from space, and certainly more plentiful than samples returned by space missions, they offer scientists a great opportunity to find out what asteroids, comets, and even other planets, are really made of. They are heavily studied on Earth for this reason.

Despite the formation of a fusion crust, the effects of atmospheric entry can be rather harsh and destructive. Those rocks with lower compressive, or lower crushing, strength are less likely to survive the experience; if an object survives deceleration through the atmosphere, then its compressive strength must be more than the maximum aerodynamic pressure it experiences. Aerodynamic pressure is directly proportional to the local density of the atmosphere, which is dependent on which planet an object encounters. So, for example, Mars has a thinner atmosphere than Earth that doesn't act to slow down incoming objects as much and explains why space engineers have to think very carefully about landing spacecraft on the red planet's surface, since their deceleration systems can't be pre-tested on Earth.

The compressive strength of a rock is controlled by its composition: its proportion of rock minerals, metals, carbonaceous material, volatile phases, amount of pore space and how well its component materials are packed together. For example, hardy space rocks, such as those from the iron-rich asteroids, tend to survive the extreme changes in temperature and pressure as they hurtle at great speed through the Earth's atmosphere. The stony meteorites are also quite robust, even when they contain little or no iron. Although iron is strong, rock minerals themselves can be very well-bonded to create a tough piece of rock, too. The meteorites that are less likely to survive atmospheric entry intact are those that contain a higher percentage of volatiles, pore space, carbonaceous phases and so-called hydrated minerals – those that have accommodated water into their growth structure. Such phases are in high abundance in the meteorites known as carbonaceous chondrites and also the comets. These objects are, therefore, more sensitive to the effects of heating and cannot withstand the aerodynamic forces they experience as they travel through Earth's atmosphere. In some cases, they are nothing more than a loosely consolidated handful of fluffy snow with some dirt mixed in. Even if you threw a snowball made of such a mix of materials you might expect it to disintegrate in the air. This demonstrates why a large sample of a comet is generally considered unlikely to survive the harsh pressure and heating effects of atmospheric entry without melting, exploding or breaking up into very tiny pieces. As such, despite the large collections of meteorites on Earth, scientists are still not certain that they have found a large meteorite specifically from a comet because of the extremely fragile structures they are expected to have. The result of all this is that some space rocks are over-represented as meteorites on Earth simply because their compositions withstand the effects of atmospheric entry better.

Where to find a meteorite

If you want to search for your own, often valuable, meteorite, then incoming space debris sometimes makes itself rather obvious by producing a blazing fireball in the sky as it heats up in the atmosphere. Such events happen all over the globe but are obviously more likely to be noticed when they occur over well-populated areas, such as the one in Chelyabinsk, Russia, in 2013. The Chelyabinsk meteor fireball was so bright that it could be seen easily in the day and it also produced a loud sonic boom as it entered the atmosphere, caused by an 'air blast' explosion that broke the rock into pieces. Even people who weren't looking up in the sky at the time were alerted to its arrival on Earth. The energy released in this air blast event was equivalent to 500 kilotonnes of TNT, or 30 times the energy released by the Hiroshima bomb. This is a sizeable blast considering the incoming asteroid was thought to be only 20m (66ft) in diameter. The meteorite debris produced by the explosion was scattered over the local area around Chelyabinsk, allowing people to find fragments of the asteroid as meteorites on the icy ground. However, it wasn't the rock itself that caused any threat to humans, as luckily no one was hit directly. Instead, the sonic boom shattered windows throughout Chelyabinsk, injuring many people with flying shards of glass. If such an impact had occurred in a sparsely populated area, or over an ocean, it would have caused little damage and few people would even have known it had happened. However, scientists would have picked up the air blast regardless, because it was large enough to be recorded on seismometers around the world, registering a force equivalent to that of a 2.7 magnitude earthquake.

Despite the 'all-seeing eyes' of the seismometers, when a meteor enters the Earth's atmosphere in a well-populated region, eyewitness reports of fireball sightings can be really useful for scientists, allowing them to calculate where the rock will have made landfall, so that they can go and collect samples. If the public have recordings of the meteor, as was the case in Chelyabinsk where the fireball was inadvertently captured on many car dashcams and closed-circuit television systems, then a trajectory can be estimated for the incoming rock. Scientists can then backtrack these trajectories to known space objects that were passing through the local Solar System neighbourhood at the time, so they might be able to connect it to an object in space.

Even without a fireball sighting, it is fairly easy to find meteorites on the Earth's surface, provided you know the right places to look. Scientists and meteorite hunters concentrate their collection efforts in largely unpopulated desert regions such as the Sahara and Antarctica. Meteorites that are at least 1cm (0.4in) in diameter, black and often shiny are easier to spot against a background of sand or ice. The lack of liquid water in these places is also a huge bonus, helping to preserve the fallen rocks, saving them from contamination by earth bugs. Liquid water is a great solvent for transporting materials to and through a rock, so the less of it a meteorite encounters, the better. Hence, the effects of terrestrial weathering are much diminished in desert regions compared with more temperate or tropical regions. That's not to say that you can't find a meteorite in a warm, vegetated area, as they land all over the globe, but they are much harder to spot, becoming buried very quickly. So, if you see a large fireball in the sky and think it's a meteorite, then you need to follow the trajectory of the incoming fireball and go out looking for the rock immediately. A meteorite that is collected soon after its arrival is called a 'fall' and one that is found many years later is a 'find'. Falls are often judged to be the best meteorites because of their limited exposure to contamination from weathering or bugs.

To focus on collecting more meteorite falls, in 1997 scientists in Australia set up the Desert Fireball Network. Although not the first of its type, this collaborative project between university researchers and an observatory at the Western Australian Museum has been very successful, involving a system of digital cameras that capture images of incoming fireballs over the Australian desert. Special software is used to calculate the speed and direction of the incoming rock to predict a potential landing zone, which is fairly well-defined because of the triangulation enabled by the network of cameras. The scientific team achieved their first success in 2007, accurately predicting the touchdown of a large incoming meteorite, now known as Bunburra Rockhole after the place it was eventually found. The complex computer systems the scientists designed can also backtrack the journey of the rock from space to calculate from which asteroid, or planetary body, the rock might have originated. This is much more accurate than relying on eyewitness reports but, obviously, this network is not available planet-wide.

If you want to collect a really large meteorite fall, then you'll need to be patient, as sizeable pieces of space rock don't impact Earth's atmosphere very often, which is surely a relief. There is an inverse relationship between the size and frequency of meteorite impact events. Incoming rocks the size of double-decker buses are, thankfully, very rare. The dinosaur-killing space rock is thought to have been around 15km (9 miles) in diameter, which doesn't sound like a huge rock in comparison with the Earth, which stands at over 12,700km (7,900 miles) in diameter, but the effects that even a rock of this size could have on our planet should not be underestimated. However, the effects of an impact don't just depend on the size of the incoming rock, they also depend on its composition. Certainly, a hardy iron asteroid impacting the Earth will cause considerable damage near the collision site, and probably for many miles around depending on its size. However, a less consolidated rock, with higher abundances of volatile material that might easily be obliterated during atmospheric entry, can still cause a great deal of damage if its disintegration produces an air blast.

In 1908, a large meteor impacted a largely unpopulated region of Tunguska in Siberia and even though no impact crater has ever been found, and no rock samples, the incident is still classified as an impact event. The fireball generated by the incoming space rock as it made its entrance was seen by people in the area but, before it made it to the ground, the meteor exploded in spectacular fashion in an air blast similar to that at Chelyabinsk, but 10 times larger. The shock wave from the blast is thought to have registered at 5.0 on the Richter scale, and it was responsible for felling some 80 million trees over a region 2,000km^2 (500,000 acres). The point is that, had such an event happened in a well-populated region, the resulting damage would have been far more significant. It is possible that the incoming meteor was of a similar composition to that at Chelyabinsk and that there were, indeed, resultant meteorites, but they were very small, making them impossible to find. Even now they would be hard to locate, being lost to the undergrowth. Alternatively, there is a possibility that Tunguska was a volatile-rich comet that left behind only dust-sized rock fragments, which were far too small to detect.

Finders keepers

Meteorites are fascinating objects for many reasons – after all they are specimens of space and, understandably, it is not only scientists who want to handle them. Part of the appeal of meteorites is their relative rarity on Earth, with rarity adding value. Many meteorites can be worth a small fortune, often more than the price of gold, with some selling on the open market for $1,000 a gram, or more. The fascination with meteorites goes back at least as far as ancient Egypt, where they were collected and fashioned into jewellery and tools. Iron-rich meteorites were found in the tomb of Tutankhamun, suggesting that they were held in very high regard. Egyptian hieroglyphs referring to meteorites translate as 'iron from heaven', signifying that the Egyptians knew where these rocks had come from and appreciated them for their high metal content.

At the present day, meteorites are still valued for their beauty, rarity and scientific importance, but one of the more complicated issues surrounding the modern acquisition of meteorites on Earth is working out who owns them. It would be great if it was simply a case of finder's keepers, but it's not that straightforward in most cases. If you want to see awe-inspiring examples of meteorites, then you would be well-advised to head to museums and universities around the world, as such institutions are fortunate enough to be the custodians of these precious samples and often display them for the public. But these organisations don't necessarily own the samples, and how the rocks made their way to them is not always simple. The good thing is that once they are lodged with learned bodies, the samples are well cared for and scientists have a way to access them for research. However, many meteorites are held in private collections around the world where they are often not available to be viewed by scientists or the public. There is, in fact, a whole economy related to the buying and selling of meteorites, with the rarer ones – generally those proven to be from other planets such as Mars, but also some rare asteroid samples – being worth the most when they change hands.

Despite the potential downside that private collections of meteorites could have on the advance of science (*i.e.* locking the samples away from the public eye), without the demand for meteorites by members of the public, it might be that many of them wouldn't be collected at all. The high market value of meteorites encourages bounty hunters worldwide to search and collect these space rocks to sell for profit. Without these people in pursuit of meteorites there would be much smaller collections available for study, as many of these 'rocks for profit' do eventually become available to the scientific community. Private collectors of meteorites are often happy to donate a portion of their sample to scientific organisations so that their rock can be classified by experts. After all, to the untrained eye a space rock can look like any other, but with the right scientific instruments it is possible to work out where it came from, and if it is proven to originate from Mars, for example, then its value may be very much inflated. When a meteorite is studied by a scientist in this way, then the classifying organisation normally receives around 20g (0.75oz) or 20 per cent of the rock to keep in their collections. However, meteorite dealers and collectors may also donate to scientists for purely philanthropic reasons.

On the whole, determining who first owns a fallen meteorite will depend on where it is found. Many modern-day meteorite collections are made in the Sahara Desert, often by nomadic meteorite hunters. These meteorite chasers have become quite skilled at finding the best samples. Don't let their nomadic description fool you – they even use four-wheel drive cars with satellite navigation to do the job. However, in the case of the Saharan meteorites traded on the open market, it's not always clear exactly where they were found, the reason being there are laws restricting the export of meteorites in some of the countries in the Saharan region. The workaround is that meteorites are smuggled into adjoining countries where the laws are not so particular. Such 'porous borders' exist from Algeria into Morocco. An advantage for the meteorite hunters of not providing an exact location for where they found their rocks is that their search sites remain secret. This is especially useful when they are collecting rocks from a strewn field – created as a meteorite breaks up on entry – as they can return to continue their search for more samples. When the Saharan meteorites are named, they are designated 'NWA' for north-west Africa, with this region representing one of the key sources of important meteorites in collections worldwide. In fact, many meteorites from Mars and the Moon come from north-west Africa, as the nomad hunters have become particularly adept at finding them. It's worth their time to be selective, since these rare rocks fetch a much higher sum on the open market than standard iron or stony meteorites.

In many countries, including the US and the UK, a meteorite is the property of the owner of the land on which it is found. The owner is free to trade the rock or keep it in a private collection, as they so wish. If the meteorite is found on public land, then it is the property of the government. In many of these cases the local or national museum to where the meteorite was found will try to obtain a sample of the rock for the public collections, which, if successful, allows the meteorite to undergo extensive documentation and be available for scientific researchers to study.

In Australia, another key region for meteorite finds (remember, dry desert regions are good), ownership laws are a bit complicated. On the whole, meteorites belong to the state in which they are found, and in some states they are then required to be kept in state museums. However, in Queensland, Victoria and New South Wales the laws are the same as the US and UK. Australia has passed laws on the export of meteorites from the country and permits are now required under the Protection of Movable Cultural Heritage Act (1986). The disadvantage of such a law is that fewer meteorites are now collected in Australia because many meteorite hunters will not work in regions where they can't export for profit – although this could be seen as a positive thing because the samples are still there for dedicated scientific expeditions to locate.

Antarctica is a very important region of the Earth for the collection of meteorites, with its main drawback being its relative inaccessibility – although this is also advantageous because it means the samples are kept safe. Many countries are actively involved in Antarctic collection efforts because of the near-perfect preservation of the rocks in the ice, even though they have often been there for many years. In some cases, the meteorites may have been there for up to 2 million years, but having essentially fallen into a deep freezer they can be considered almost 'new', or like a fresh fall. The other advantage of Antarctica is that the glacial flows concentrate the meteorites, and they're relatively easy to spot – black rocks in a background of white ice. If meteorites fall on the high Antarctic mountains then they are rapidly covered by snow and the ice flows downhill, carrying with it its embedded space rocks. The ice may come up against a mountain range, such as the Transantarctic mountains, where it is gradually eroded away to reveal its bounty of black rocks.

Going as far back as 1969, it is the Japanese who first began searching for meteorites in the Antarctic ice. Since 1976 ANSMET (Antarctic Search for Meteorites), an American-led organisation, began collection efforts but there are many other countries involved, too. Each year scientists set out on expeditions for a couple of months in the local summer to collect meteorite specimens to be returned to laboratories for scientific examination. The numbers of samples collected by each organisation is huge – for example, ANSMET have collected over 16,000 meteorites from the region in 40 years and these meteorites become available to the scientific community worldwide for analysis. Arguably the most famous Antarctic meteorite is ALHA84001 (ALH representing the Allan Hills, where it was collected in 1984), which is thought to be from Mars. It was also the centre of a controversy in 1996 when some researchers announced that they'd found evidence for life inside the rock – a peculiar make-up of carbon and magnetite produced by some bacteria on Earth, the presence of which might have signalled biotic production. It is now thought that these ostensible microfossils were either contamination or abiotic structures that simply resembled the shape of bacteria, but the meteorite made a name for itself in the process.

The large collections of meteorites that we have here on Earth, wherever they are held, were collected from all over the world and have mostly been documented and studied in detail over many years. In fact, when we have large enough samples, they can be reanalysed many times over the years as scientific instruments improve, thus allowing for even more refined measurements. Even though there are many kilograms of space rocks here on Earth, care must be taken with these free cosmic samples, because some of the rocks in these collections are very rare. It is therefore important that these samples are carefully stored and preserved so that they are still available for future generations of scientists to study in their even more advanced laboratories.

Figuring out where meteorites came from

As we've seen, meteorites are not always from asteroids; sometimes they can be pieces of comets or other planetary bodies, such as Mars or the Moon, that have been ejected into space and ended up travelling to Earth. However, larger samples of these space objects are extremely rare on Earth, and although scientists are certain that they've collected Martian and lunar meteorites from the Earth's surface, they are yet to identify a meteorite that originated from one of the other rocky planets – Mercury or Venus. This may seem surprising. After all, these planets are also some of our closest neighbours in the Solar System, so you might expect samples of them to be more common on Earth. There is always the possibility that a sample of one of these planets is hiding inconspicuously in Earth's collections, having been catalogued incorrectly. The reason why scientists might not realise they have a piece of Mercury or Venus is that, so far, no spacecraft have visited these planets to collect rock samples. Therefore, recognising samples of them as meteorites on Earth is difficult as no one knows what to expect, with nothing to compare them to. Conversely, hundreds of kilograms of lunar rocks have been collected from the surface of the Moon and analysed on Earth, giving scientists a direct comparison with meteorites they suspect to have originated from the Moon.

From this, you may start to question how scientists can be so certain they have samples of Martian meteorites in their collections, since no samples have yet been collected from the surface of Mars and returned to Earth. In spite of this, the surface of Mars, and its atmosphere, have been extensively studied for decades with scientific instruments on landers and orbiters during successful space missions. These measurements of the red planet have, among other things, provided the composition of Martian atmospheric gases. The most recent of measurements was made by the SAM (Sample Analysis at Mars) instrument on NASA's *Curiosity* rover, but other measurements were made by both the *Viking* and the *Phoenix* landers dating back to the 1970s. Atmospheric measurements on Mars provide invaluable information, not only for understanding the history of the red planet, but also to provide a comparison with the rocks on Earth that are thought to have originated from Mars. Meteorites that are suspected to have been knocked off Mars bring with them Martian atmospheric gases they trapped within their rock structure when they formed. When these samples are cracked open in laboratories on Earth, the gases trapped inside can be measured and their composition and abundance compared with the measurements made in space – like a sort of planet 'fingerprinting'.

Such gas measurements involve complex scientific instrumentation and, in general, are only undertaken after a range of simpler techniques have been used first, in the hope that these can help to reveal information about the provenance of the rock. For example, it is usually simple for a trained lunar scientist to tell a terrestrial from a lunar rock by studying the composition of one of the main rock-forming minerals, plagioclase, within the sample. This involves relatively low-cost instrumentation – a microscope – and doesn't take a long time to complete. In many cases, micro-analysis techniques, such as looking at a thin section of the rock, are the starting point for the classification of meteorites and it is only if these fail to demonstrate where a meteorite comes from that scientists start to utilise the more complex laboratory techniques.

When the simpler techniques fail, one of the key diagnostic tools scientists use to work out a meteorite's provenance, and one that can be used on every type of rock from space or Earth, is to analyse the abundance of isotopes of oxygen within the rock. Isotopes are quite simply versions of the same element that have different numbers of neutrons within their nuclei. Even though this results in a very small difference between the isotopes, it can still be very useful to scientists. I'll explain why now.

Oxygen is one of the most abundant elements in the Solar System and is a key constituent of rocks. There are three isotopes of oxygen, we breathe all of them, their only difference being their slightly different atomic masses due to their varying numbers of neutrons. Over 99 per cent of Earth's oxygen is oxygen-16, so-called because each atom contains eight protons and eight neutrons. There are, however, also small quantities of the heavier oxygen isotopes – oxygen-17, which has one extra neutron and represents about 1 in 2,000 atoms of oxygen; and oxygen-18, which has two extra neutrons and is found in 1 in 500 atoms of oxygen. The important thing is that other planets, asteroids and comets have their own unique ratios of these minor isotopes, giving them a distinctive oxygen isotope signature compared with Earth. This happens because the isotope ratios are affected by the exact conditions under which they were incorporated into the rock. In a nutshell, if two objects have a different oxygen isotope composition, then they formed in separate locations, or times, in the Solar System. So, if a rock has the oxygen isotope composition of Mars, then it is definitely from Mars and not, for example, from Venus, which is expected to have its own composition. As an aside, the composition of Venus is, as yet, undetermined as there are no samples that definitely originated from Venus on Earth. As you'll see, analysis of oxygen isotopes gives scientists a convenient way, once again, to 'fingerprint' different planets, asteroids and comets, allowing them to figure out if objects are related and where they formed.

Rather excitingly, the oxygen isotope composition of the Sun has been determined, as it was measured by the NASA *Genesis* mission in 2011, which captured samples of the solar wind to return to Earth for analysis. The mass of the Sun is 99.8 per cent of the Solar System, so it can be considered as the 'bulk Solar System' value. Anyway, it happens to be characterised by a relative depletion in the oxygen-17 and oxygen-18 compared with Earth and the other planetary objects. What is known about the other planets is that their bulk oxygen isotope compositions have a rather restricted range, *i.e.* one planet is not very different from another. The reason is straightforward: they all formed within the inner Solar System, at around the same time as each other. The chondrites – meteorites that are the 'building blocks of planets' – share a similar range to the rocky inner planets, having formed in roughly the same location and at around the same time.

Consequently, determining the oxygen isotope composition of a meteorite can help scientists to figure out where it originated. This is useful because it is not always easy to work out if two meteorites are from the same parent body by just studying them under a microscope. After all, two humans can look very alike but, unless they are identical twins, their fingerprints would show they were different. Isotopic fingerprinting of meteorites is an often useful step in the classification process. In addition, even without launching sample return missions to collect rocks from Mars, scientists have been able to establish the oxygen isotope composition of the red planet because, in the same meteorite rocks in which they've measured Martian atmospheric gases to compare with the spacecraft measurements, they've also been able to measure oxygen isotopes. This is a neat consequence that has meant scientists now understand quite a bit about how, and where, Mars formed, without having collected samples from the planet for return to Earth yet.

Forming the Earth's Moon

Using oxygen isotope studies of lunar rocks collected by space missions or those that have arrived as meteorites, scientists have been able to investigate details about the formation of the

Moon. Have you ever stared into the night sky and wondered how the Moon got there? Maybe you take it for granted and just assume it's been there for as long as the Earth. But it hasn't. The Moon has an interesting story to tell. There is a strong international consensus that the Earth and the Moon are identical in terms of oxygen isotopes. This information has helped scientists to work out that the Moon certainly can't have formed in a far-flung part of the Solar System, otherwise it couldn't share the same oxygen isotope composition as Earth. This means that it is highly unlikely that the Earth's Moon formed as a 'captured' Moon, as is the case with some of the satellites around other space objects where the gravity of the larger body acts to seize a nearby smaller object. A captured Moon would be expected to have a different oxygen isotope composition, even if only slightly, to the object it is detained by.

The composition of the Earth and Moon is so close that it led scientists to propose that the Moon must have formed directly from the Earth itself. Current models suggest it formed during a large-energy impact early in Earth history that caused part of the Earth to be ejected into space as impact debris, until it eventually coalesced into one body and then cooled to form the Moon. The object that collided with Earth is suggested to have been near enough the size of Mars, which is a scary thought. But without large impacts of space objects on Earth we would never have got our Moon. Without the Moon, our world would be a very different place. Our nights would be darker, our days would be shorter, our tides would be lower and we'd probably experience some even more wild weather. So, let's be thankful for this significant impact in Earth's history.

The Moon's story is highlighted here, in particular, because it is one of the planetary bodies that scientists have, arguably, spent the most time analysing – picking apart its history by measuring a large quantity of its rocks. Yet, despite all that has been learnt about the Moon – and that's a lot of information – scientists don't agree on every aspect of its formation. There is still work to do, but it is really just dealing with the finer details. This just goes to show that understanding the origins of the countless different types of comets and asteroids is a major challenge for space scientists and that a great deal of detailed scientific investigation is going to be required for them to complete the task.

What's in a comet or asteroid?

To probe into the very earliest years of Solar System history, scientists will primarily look to sample comets, as they contain the most primitive ingredients from the outer disc of the solar nebula, which they've preserved impeccably well over time. However, the more evolved, rocky asteroids also have their uses, recording information about the inner disc region of the early solar nebula. This is another important place to journey to if you want to piece together the history of the planets and Solar System as a whole.

The objects that now inhabit the inner and middle regions of the main asteroid belt sample a different, slightly later phase of Solar System formation to the comets – albeit one that is still very close to the Solar System's birth, but that also overlaps with the formation of the planets. As we learnt throughout Chapter 3, instead of the primitive fluffy dust collected by the comets, the asteroids that formed within the inner disc were made of brand new materials that came into being a little later within the Solar System itself. These were formed by the infant Sun when its heat transformed solar nebula dust into special high-temperature mineral phases. Some of these new materials were small, sub-millimetre to centimetre-sized components that are found in meteorites. These components are the chondrules, after the Greek word for 'seed' or 'droplet', and the calcium-aluminium-rich inclusions (CAIs), and they are very valuable to scientists as they formed within just a few million years of the birth of the Solar System.

Chondrules make themselves rather obvious in rock samples, despite their small size, by the fact that they are spherical objects. They are thought to have been molten droplets during the phase of planet building, and are formed of key rock-forming minerals such as olivine and pyroxene. The oxygen isotope compositions for individual chondrules have been measured in the laboratory – after plucking them from the rock for analysis – and the values they produce are similar to those of the rocky planets. This evidence is used to imply a genetic link between chondrules and planets, suggesting that the chondrules are the raw ingredients for the planets, having formed in roughly the same region.

CAIs, on the other hand, can form a range of interesting and beautiful shapes, from irregular to perfectly spherical, with the more unevenly shaped CAIs being made of an intricate mixture of different minerals. The very high formation temperatures calculated for these minerals indicate they were created even earlier than the chondrules, but perhaps by only a million years or so, which is not much on the grand Solar System scale. These high-temperature minerals have names such as corundum and hibonite, containing an abundance of calcium and aluminium, hence the name for the objects they are contained within. The CAIs have a much wider range in oxygen isotope compositions than the chondrules, ranging from the composition of the Sun to that of the planets. This range reflects the fact that these small components formed over a period of time, albeit a short one in relation to total Solar System history, and in a range of different places. Some CAIs formed from the early solar nebula dust existing right next to the Sun very soon after it formed; they recorded within their rock minerals the same oxygen isotope composition as the Sun at the time. Other CAIs formed a little later on (but here we're talking less than a million years later), but before the formation of the planets, possibly a bit further away from the Sun, so they inherited a different oxygen isotope signature that was slightly more 'planetary like'.

The age of the Earth

A technique known as radiometric dating has allowed scientists to pinpoint the exact time that many of these important small Solar System objects formed, their 'formation' simply being the moment they solidified as rocky minerals. This has, in turn, allowed scientists to come up with an accurate estimate for the age of the Solar System itself, since these are the first solids to have formed in it and, therefore, the oldest objects scientists can measure. In addition, radiometric dating has given scientists the information they needed to reconstruct a timeline of events in the Solar System, placing timescales on how long objects took to form, and how long it took for everything to get to where it is today.

The techniques used to date the CAIs and chondrules are not dissimilar to those of carbon dating, which most people have heard about in relation to archaeological artefacts and crime-scene investigations. The problem with carbon dating is that it can only be used on materials, including rocks, that formed a maximum of 60,000 years ago – the blink of an eye in relation to the age of the Earth. The reason for this is that the half-life for the decay of carbon-14, the radioactive isotope of carbon used in carbon dating, is only just under 6,000 years, so it decays too quickly to make it useful for dating rocks that are billions of years old. However, there are many elements, such as uranium-235, which has a half-life of 704 million years, and potassium-40, with a half-life of 1.25 billion years, among others, that can be used for dating these more ancient rocks. A range of these precise isotope dating systems have been applied to the CAIs to give an age of about 4.6 billion years. These are the oldest ages recorded for any Solar System materials, in keeping with experiments suggesting that the high temperature mineral phases contained within CAIs are the first solids to condense in the solar nebula. So, the CAI ages are used to indicate the age of the Solar System itself. This time is often referred to as time zero ($t = 0$), which may be just a few tens of thousands of years after the actual birth of the Solar System. It is only possible to measure 'time zero' by analysing solid material, which presumably formed a little after the 'real' time zero.

The rocky planets are thought to have originally contained all these same ingredients – chondrules and CAIs – having been built up from the asteroids themselves. However, the hefty size of the planets meant that they experienced extensive geological re-processing, melting and re-melting in the intervening years since they formed. The planets have essentially cooked their original solar nebula ingredients, transforming them into other mineral and rock assemblages that obscure the early Solar System history their raw ingredients held. It's a bit like the ingredients for a cake, which are then mixed and baked. The end product looks very different to the pile of flour, sugar and eggs the baker started out with. Most people would struggle to identify a cake's recipe with only the final baked product to work with. They might guess at the majority of the ingredients that were added, but they probably couldn't work out the exact quantities. After all, most people follow a recipe to bake a cake, with a list of the ingredients that are needed and in which quantities to add them to achieve a tasty concoction. With a planet, all that's left is the end product, the cake, which is obviously great but it would be useful to know how it was produced if you liked it. Only then can scientists start to understand what the conditions were like early in Earth history at important times such as when life began.

Altering meteorites in space

Despite all this, not all meteorites can be used in the same way to answer the same questions, and scientists must be picky about which meteorites they study if they want to probe back to the Solar System's very earliest years. The reason is that some of the asteroidal meteorites haven't perfectly preserved their early-collected materials. Without question, many of the asteroids spent a great deal of time near to the Sun after they gathered together their precious Solar System solids. The amount of subsequent heating the asteroids suffered didn't, in many cases, completely destroy their components, but it could result in some changes to their chemical make-up and structure. Such 'alteration' is dependent on the amount of water, organic matter and other volatile materials they contained. Alteration could be thermal (from heat) or aqueous (from water), or a mix of the two. Many asteroids suffered some alteration from shock, too. They were right in the thick of it, so to speak, located in the crazy place that was the early inner Solar System, and they underwent collisions soon after they formed, bumping into one another as they journeyed around the young Sun. The effect of these impacts was that the components they contained were heated and pressurised, bringing about further changes to their original structure and chemical composition.

The good thing is that after many years of studying meteorites these are processes that scientists, on the whole, understand very well. In many cases, scientists can unravel the effects of the temperature and pressure changes experienced by the rocks to figure out the history of the parent asteroid in

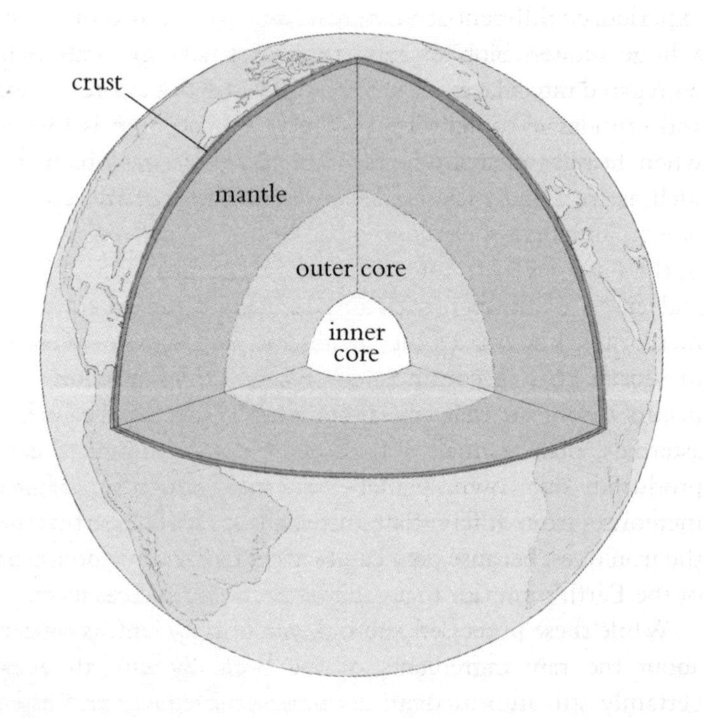

Figure 5 *The interior of Earth. A cross-section of the planet showing its differentiated, layered structure of a solid inner core, a liquid outer core, a mantle and a thin crust.*

space. The effects of alteration are seen as useful in some ways, as they allow scientists to understand more about the events that occurred in the inner disc soon after the asteroids formed. Even so, if scientists want to journey back to investigate the Solar System's oldest solids, then it is better for them to focus on asteroids that experienced little or no alteration after they formed. In this respect, it is the stony meteorites, otherwise known as the chondrites, that are particularly important because they are only thought to have experienced temperatures and pressures that were sufficient to bind their collected rock components together as a single body, without totally obliterating their fine-scale structures and inherited history.

Having said that, even the highly processed iron meteorites are useful for piecing together events in the early Solar System. Such meteorites originate from the core of asteroids that experienced differentiation: where the object was at one point a huge molten blob of rock that internally gravitationally segregated into a layered structure consisting of a core, mantle and crust as it cooled down. The iron-rich core is formed when heavier elements present in the molten blob, metals such as iron and nickel, sink to the middle of the cooling object. The lighter elements, such as silicon and oxygen, stay in the outer layers to form a mantle and a crust. The mantles and crusts of differentiated asteroids are represented in the meteorites known as the achondrites – a type of stony meteorite that is common on Earth. Differentiation only occurs in objects that are large enough, such as the heftiest asteroids, planetesimals and planets, and is known to have produced our own planet's internal structure. Hence, meteorites from differentiated asteroids are useful, particularly the iron ones, because they can be used to find out about parts of the Earth's interior that can't be accessed, such as its core.

While these processed asteroids can't help scientists to learn about the raw ingredients of the Solar System, they can certainly still inform them about the subsequent processing that occurred once the ingredients were assembled into large objects like our own planet. The result is that nearly every meteorite sample, whether it came from an object that formed very early on, or one that formed later and underwent extensive alteration next to the Sun, can be useful to scientists. The beauty of studying comets and asteroids is that they sample not only slightly different, but very early, stages of Solar System formation, but also different regions of the early protoplanetary disc. Sampling them also means scientists avoid having to unravel 4.6 billion years of complex planetary processing.

Uniting astronomy and sample science

If planetary scientists could time travel, they might choose to go back exactly 4.6 billion years to see all the early Solar System action in progress. However, apart from studying the products of this phase of Solar System formation, there's another way to 'see' this period of time. For this, scientists have to look outside of our Solar System, at other distant planetary systems going through their own early phase of formation. After all, star and planet formation is something that is happening continuously all over the Universe. Although humans can't physically travel to these other stars and planets to sample them directly, they can certainly look at them through telescopes, obtaining a bird's eye view of these other systems and comparing them with our own. One of the best telescopes for this task is ALMA, the Atacama Large Millimeter/submillimeter Array, which is actually a single telescope composed of 66 high-precision antennas located deep in the Atacama Desert of Chile. ALMA has glimpsed many distant star and planetary systems, capturing images of them at different stages of their formation. Although this telescope can't resolve down to the fine-scale level of detail of the formation of CAIs and chondrules, it can still show scientists when, and where, the planets start forming on a grander scale, tracing the location of hot and cold materials in young protoplanetary discs. These 'real-time' observations are used by astronomers and planetary scientists to hone their ideas about Solar System formation, with the Grand-Tack and Nice models, as discussed in Chapter 2, for example, incorporating astronomical data as well as laboratory analyses of Solar System materials.

Astronomical data is also useful for scientists studying meteorites in the laboratory. After all, meteorites are the offspring of asteroids and comets, and they have journeyed from being objects in space to landing on Earth. However, with so many asteroids out there, and a very large and varied meteorite collection on Earth, linking them is not always easy. Firstly, scientists often look for ways to categorise asteroids and meteorites based on their general characteristics and compositions. If a meteorite can be studied on Earth and its characteristics compared favourably with remote measurements of asteroids made by telescopes, then scientists can start to say from which asteroid, or group of asteroids, a particular meteorite might have originated. To achieve this, asteroids are studied remotely with telescopes and space missions, and in the laboratory with microscopes and mass spectrometers to find common links in the data obtained by different techniques. In some cases, the same techniques can be applied by telescopes on asteroids in space and by laboratory instruments on little pieces of meteorites back on Earth, allowing for as close a direct comparison of these objects as is possible.

As we've seen, astronomers have been able to divide the asteroid belt according to the composition of the asteroids it contains in different locations, with the inner belt asteroids being essentially water-free and composed of more iron and rock, and the outer belt asteroids being rich in water and carbon. These ideas have been shaped using spectroscopic measurements to enable scientists to understand the colours of the asteroids to determine which minerals and volatiles they contain, and, looking at their albedo to see how reflective they are, to estimate if they contain lots of ice (high albedo) or carbon (low albedo). However, most of the time these remote telescope techniques can only look at the outer layer of the asteroid, the surface exposed to space, and they can't probe into the depths of the space rock. This is another reason why measurements made on meteorites in the laboratory are useful, and complementary, as they are samples of the inside of these objects and can help to further reveal how the asteroids formed.

What scientists have learnt is that the iron meteorites are almost certainly sourced from the asteroids located in the inner parts of the main belt, the M-type iron-rich asteroids that have lost their crust, mantle and volatiles. These asteroids are more evolved objects that have undergone melting and differentiation, and suffered violent collisions both before and after differentiation that acted to destroy their early-inherited primordial ingredients, or to peel off their outer layers to leave behind just the hardy inner core. The carbonaceous chondrites, on the other hand, are most likely sourced from the group of asteroids located in the outer main belt, those objects that are most like the comets, containing early-inherited Solar System ingredients in pristine, or near pristine, form. In some cases, it is possible to pinpoint the exact asteroid, or family of related asteroids, that a meteorite originated from, but this is generally only possible for the larger space objects, as these are the ones that can be viewed remotely with telescopes.

One example is the asteroid Vesta, the second largest body in the main belt after the dwarf planet Ceres. There had always been a suspicion that a certain group of meteorites on Earth made up of rocks named the Howardites, Eucrites and Diogenites, or HEDs for short, resembled Vesta. They appeared to have originated from a large differentiated body and their compositions were similar to what was known of the surface of Vesta as measured remotely by telescope. However, it seemed there was a problem because Vesta is quite far away from Earth and located in an unfavourable position within the asteroid belt to supply meteorites to Earth, whereas, the HED meteorites represent about 6 per cent of all the meteorites that arrive on Earth, a fairly large proportion. The HED meteorites are more common than meteorites from the Moon and Mars combined, both of which are much closer to Earth. So, the conundrum was that, despite the fact that HED meteorites seemed to resemble Vesta, it was hard for scientists to be sure they were 'one and the same' because they couldn't fully explain how the rocks got from there to here. This conundrum, in part, inspired the *Dawn* mission, which was able to help by taking a much closer look at a large region on Vesta called the Rheasilvia Basin. This contains two craters and an astoundingly large central peak rising to 22km (13.6 miles) – twice the height of Mount Everest. Scientists thought that this significant hole in the side of Vesta must have been created by a cataclysmic impact that would have resulted in the displacement of a great deal of debris into space. It is this debris, some of which would have been the size of small asteroids, that is thought to account for the supply of the HED meteorites to Earth. This couldn't be confirmed without taking a closer look at Vesta.

The *Dawn* mission studied the Rheasilvia Basin in detail, taking many remote measurements of its rocks to compare with studies of HED meteorites. The result was that scientists concluded that HED meteorites were the same composition as the rocks in the Rheasilvia Basin. It's thought that the rocks that were violently ejected from the basin – now known as the Vestoids – ended up making their way to a special gap in the asteroid belt whose location can result in objects being thrown more easily into an Earth-crossing orbit. This explains why the HEDs (or Vestoids) are so common as meteorites on Earth. Although there's probably still much to learn about this fascinating large asteroid, it is certainly helpful that scientists have realised they can study the HED meteorites on Earth to understand more about Vesta's history, heightening the scientific value of the HED meteorite group.

As time goes on, existing and new meteorites will continue to be studied in detail in laboratories the world over, allowing scientists to learn much about the formation and structure of the Solar System. However, as more asteroids and comets are studied in detail with telescopes, and up close with dedicated space missions, further links like those between Vesta and the HEDs can be made. Such complementary measurements and resultant associations between space and Earth-based rocks will make it easier for scientists to piece together the intricate structure of the Solar System, understanding how and where everything formed – from the largest of the planets such as Jupiter to the smallest of the grains such as CAIs, chondrules and stardust.

CHAPTER 4

Shooting Stars and Space Dust

Shooting stars have fascinated me since a very young age, but when I was little I rarely had the patience to sit still for long enough to see any. The idea that we can 'make a wish on a shooting star' always excited me, and I always had a wish all ready to go but, most of the time, I failed to see a single star shooting through the sky. As we grow older, and learn to have more patience, many of us still think about making that wish, but we also learn that it's not sufficient to just look up at the sky occasionally, because often the sky we look up at is not dark enough. Throughout the studies for my geology degrees I learnt that there are places in the world where you are almost guaranteed to see a shooting star if you stare at a patch of sky for a little while – the wilds of Scotland, the glaciers of Iceland and the outback of South Africa are a few places that I've been lucky enough to visit and gaze into the night sky, in awe of its beauty. Clearly, to improve our chances of spotting a shooting star it helps to be somewhere very dark, where little light from a nearby town or city can intrude.

Having grown up in the south-east of England, near the hub and glow of London, I really didn't appreciate how much light pollution there was until I travelled further afield. Now when I get to a place that is truly dark I always look up at night to see if I can spot one of these ephemeral yet beautiful cosmic events. Knowing that a shooting star is simply produced by a little piece of rock burning up in the atmosphere may not sound very romantic, but it doesn't diminish their appeal for me – it actually increases it. These pieces of rock dust have made their journey to Earth from places far away from us, sometimes from the very outer edge of our Solar System. In fact, what happens when we see a shooting star is that a tiny part of our 4.6-billion-year Solar System history is being destroyed in a flash of light. However, in many cases, incoming pieces of space rock are not destroyed (we'll find out why in a minute), and this is obviously good news for scientists like myself who want to study these tiny samples of the solar nebula.

Dust in space

Dust in space is known as cosmic dust, and it can be divided into groups based on astronomical location: intergalactic dust, interstellar dust, circumplanetary dust and interplanetary dust. It is the last of these that produces the zodiacal cloud, a faint glow in the sky after sunset in spring and before sunrise in autumn caused by sunlight being scattered by the interplanetary dust that surrounds Earth.

There is actually so much dust in the voids of the inner Solar System, much of it left behind by asteroid collisions and passing comets, that if you collected it all together and compressed it a bit it would form a moon 25km (15 miles) in diameter. This dispersed dust is pulled around the Solar System by gravitational forces, and the Earth can't help running into some of it as it makes its journey around the Sun. When such dust falls to Earth these free cosmic gifts present one of the simplest and cheapest ways to sample space rocks, albeit very tiny ones, without ever having to leave the safe confines of our atmosphere.

Space dust on Earth

The result of all this is that we have plenty of cosmic dust here on Earth. The problem is that most of the time we just can't see it. When you vacuum your house, you collect minute pieces of space dust along with all the background house dust, and it even gets in your hair and clothes when you go outside. We already learnt in Chapter 3 that an astonishing amount of rocky space material falls onto our planet each year – some 40,000 to 80,000 tonnes. This seems like quite a wide-ranging estimate and that's because no single technique can observe particles over the entire mass range required: from dust to boulder-sized. Most studies, however, do agree that dust-sized particles – just one-tenth to one-hundredth of the width of a human hair – dominate this incoming mass. This may sound like an alarmingly huge amount of material, but compared with the overall mass of the Earth – nearly 6 sextillion tonnes (that is 5,972,000,000,000,000,000,000 tonnes) – the amount gained from cosmic dust is negligible. The Earth loses around 50,000 tonnes of mass each year through hydrogen and helium escape to space, so it, rather neatly, has very little resultant net gain or loss of mass.

Knowing that the Earth, and probably you and your belongings, are covered in space dust may help to add a glamorous sparkle to your chores, but you'd be hard pushed to see this cosmic dust in your house should you want to collect it. The search would be like a supercharged version of 'a needle in a haystack', even with the most powerful microscopes, as the dust is too tiny to see with the naked eye and is mixed in with the background dirt and grime in your house – mostly skin cells, the shed skins of dust mites and assorted fibres. All of which are less glamorous than a sample of our Solar System.

Sure enough, by adulthood most of us have seen cosmic dust in the night sky as a shooting star, and we can increase our chances of spotting it if we look at the sky during a meteor shower. Such a celestial event occurs when the Earth passes through the trail of dust left behind by a passing comet, which greatly increases the flux of dust falling to Earth. One of the best meteor showers that happens every August is the Perseids, which can produce dozens of shooting stars every hour. Although it may not be possible to catch this dust by sitting in your back garden and staring at the night sky, you can certainly enjoy its ephemeral beauty as it shoots towards the Earth.

Not all of the dust that enters Earth's atmosphere produces a shooting star, destroying itself in the process. Many of the smallest pieces of cosmic dust enter Earth more discreetly, being better at surviving the effects of atmospheric entry and thus not producing a shooting star. Heat conductivity through the smaller pieces of dust is thought to be fast enough for the particle to be isothermal. That is, the temperature increase through frictional heating in the atmosphere is balanced by the temperature loss through radiation, so nothing happens to it. The presence of water and organic material in cosmic dust may also help to dissipate the frictional build-up of heat on contact with the atmosphere, as these components vaporise at a lower temperature than the rock components, causing them to be lost while the rock remains intact.

Particles of cosmic dust on Earth are micrometeorites (small meteorites), the very smallest of which are otherwise known as interplanetary dust particles (IDPs). These small dust grains are like the seeds of the Solar System and they originate from a wide range of Solar System objects, including comets and asteroids, many more than scientists could hope to approach with spacecraft in human lifetimes. When individual IDPs are released from their asteroidal or cometary parents, their orbits are affected by Poynting-Robertson drag – resulting from the effects of solar radiation acting on small particles up to 1mm in diameter – that causes nearly all of them to eventually reach Earth-crossing orbits irrespective of the location of their parent in space.

This means that Earth-based scientists can collect dust from space to analyse in their laboratories, something that they do regularly and which we'll learn more about shortly. But firstly, it's important to understand *why* they collect this dust. For the comet enthusiast, it's good news, as it turns out that more than 70 per cent of the dust-sized material arriving on Earth from space is thought to originate from Jupiter-family comets, while the rest is from other comets and asteroids. Of course, any rock from space is of great interest to space scientists, but dust originating from a comet has special benefits.

As we've seen, comets are the oldest objects in the Solar System, formed before the planets and asteroids, and they were plucked from the swirling cloud of gas and dust – the early solar nebula – from which our Sun and entire Solar

System formed. This means that comets contain all the earliest ingredients, including water and organic material, that went on to form the rest of the Solar System, including us. If scientists want to work out how life started on Earth, and where all of our planet's water came from, then comets (and primitive asteroids) are a good place to start looking. After all, they might have been responsible for seeding our young Earth with the necessary materials for life to take hold.

Collecting dust from space ... on Earth

One of the best ways for scientists to sample IDPs is to catch them before they ever reach the ground. Interestingly, though, it isn't necessary to go all the way into space for this. Instead, scientists sample them high in the Earth's stratosphere. The reason they often choose to collect at high altitude, rather than waiting for the dust particles to reach the Earth's surface, is the lower concentration of terrestrial dust up high, meaning that the majority of the samples collected are likely to be extraterrestrial. The early stratospheric dust collection tests carried out in 1970 were not very sophisticated, using high-flying balloons called 'Vacuum Monsters' that sucked in air and filtered the particles out. However, it was soon found that these balloon collectors became contaminated on their ascent and descent when they had to traverse the relatively dirty lower atmosphere. So, in recent decades NASA has used a research version of the U2 jet-powered spy plane, called the *ER-2*. This aircraft is designed to fly at extremely high altitudes, virtually to the edge of space at 20km (65,000ft), above 99 per cent of the Earth's atmosphere, to capture some of this precious space cargo.

The way the *ER-2* avoids sampling the relatively contaminated lower-altitude air is that the pilot doesn't open up the collection pods that are attached to the underside of the wings until they've reached the desired altitude. The collector pods are packed in a super-clean laboratory environment prior to flight and are made up of a flat collector plate covered with silicone oil that acts a bit like sticky flytrap paper except that, in this case, the flies are bits of space dust. At altitude, the collectors are exposed for a period of several hours, with cosmic dust impacting the collector with an inertial impact of more than 200m/s (450mph). The doors of the pod are closed prior to the aircraft's descent and the collectors are removed after landing to be transported back to clean laboratories.

These laboratories, housed at the NASA Johnson Space Center in Houston, Texas, are specially designed to handle such samples, and the scientists dress much like forensic officers entering a crime scene, wearing face masks, gloves, overshoes and overalls. This is not to protect them from anything ominous in the extraterrestrial dust, but to protect the collected dust from Earthly contamination. The air in these laboratories is highly purified to achieve extremely low background dust levels. This means that any contaminants found on the collector will have originated from high in the stratosphere and are either terrestrial dust, such as that powerfully ejected from volcanoes, or more commonly spent rocket fuel, which appears as rounded blobs of solidified aluminium.

The particles are removed skilfully from the collectors – after all, we're talking about rock particles that are extremely tiny and often very fragile – using a fine needle attached to a microscope, and requiring a very steady hand. The very small size of the IDPs means they are attracted to the tip of the needle due to electrostatic attraction once the needle is moved close to them. They are placed onto a glass slide and carefully rinsed with a pure solvent to clean away any adhering silicone oil left over from the collector. Despite all the effort, even at this stage it's almost impossible to know the exact provenance of the dust, as it is not instantly obvious if it's extraterrestrial or not. Even during initial observations of the collected dust under the microscope, terrestrial particles are not always easy to distinguish from true IDPs, particularly terrestrial volcanic dust – it's all just rock, after all. Because of this, the particles undergo careful screening by NASA scientists before being sent out to scientists around the world for analysis. Once scientists receive their samples they are always on the lookout for terrestrial particles that might still be masquerading as space dust in their allocation. Sometimes the only way to tell them apart is by analysing the rock dust in detail, observing the structure of the particle, the minerals it contains and its chemical composition. Only then can a scientist make an informed decision on whether they have a piece of space dust or not.

Comets and asteroids in the laboratory

Clearly, the downside of IDP collections is that, even once terrestrial particles have been screened out, it's not possible to know exactly where each particle of dust originated. It would take some work to backtrack the journey of any individual IDP to its parent in space. In fact, that is probably impossible to do, as much of the dust doesn't arrive on Earth directly from its parent body: it might circle around in space for thousands of years before settling here. So, unlike the meteorites we learnt about in Chapter 3 that can be traced by a network of cameras in the Australian desert back to an object in space, it is impossible to be 100 per cent sure where any piece of dust originated. Having said that, on at least one occasion NASA has made a dedicated *ER-2* collection when the dusty debris left behind by a passing comet was due to collide with Earth, increasing the overall flux of cosmic dust to Earth at the time. In this case, although the background 'normal' dust was still falling, it is thought that the majority of the IDPs on the collectors were from the passing comet, which was useful for scientists investigating these grains.

Despite the drawbacks – not knowing from which exact space object the dust originates, comet or asteroid – IDPs provide a valuable resource for scientists investigating the Solar System because they currently represent the main inventory of comet samples on Earth. IDP collections allow scientists to analyse a wide range of comets in the Solar System without having to send a spacecraft all the way to the Kuiper Belt or Oort Cloud to collect samples, something that would be very time-consuming and expensive.

Nevertheless, if scientists want to specifically analyse the pieces of comets in these dust collections, then they need a way to tell apart the comets and the asteroids in the laboratory. To do this, it is first useful to understand what the difference is between these objects in space. As discussed in Chapter 2, classic Solar System formation models place the comet formation zone in the far outer reaches of the Solar System, away from the heat of the Sun. As a result, comets are expected to contain minerals and materials formed at the very low temperatures found in this region. It just so happens that these cold materials are the purest, and oldest, samples of the early solar nebula that eventually coalesced to form our entire Solar System. As such, if a meteorite or IDP is found to contain an abundance of rocky materials and minerals formed at high temperature, then it can be assumed that it originated from either a planet or an asteroid – something that formed closer to the Sun. However, we do have to be a little careful with this assumption, as we'll learn in Chapter 7, but for now we'll stick with it.

Luckily, in terms of IDP collections, statistics are on our side. As we've seen, dynamical models predict that much of the dust entering Earth's stratosphere originates from comets. So, there's a high chance that any individual particle collected in the stratosphere is of true cometary origin in the first place. Of course, scientists would struggle to say whether a particle was from a comet or a primitive outer belt asteroid and, in the end, it sort of doesn't matter either way. The word 'cometary' becomes just a label and it might be better to think of all the outer belt asteroids and comets simply as 'primitive' Solar System objects instead. They are all expected to contain the most ancient and pristine solar nebula materials and, as such, are all of use when studying the formation of the Solar System. The only difference between these objects is how they have preserved their bounty of ancient solar nebula ingredients to the present day, with those objects that stayed far from the Sun probably doing the best job.

When an IDP enters a scientist's laboratory for analysis, one of the first things they will look at is its structure, or 'texture', using a high-powered microscope. As we've seen, comets, and the dust originating from them, can be described as 'fluffy' in texture because the tiny solar nebula silicate dust particles that compose the dust grain are not compacted together very well, giving the rock a high porosity (a bit like the holes in a sponge). Picking up a big piece of comet, if it were possible on Earth, would be a bit like handling a crumbly, sandy soil. Remember that we think of comets almost like a delicate little scoop of the solar nebula. Consequently, cometary IDPs are very fragile, being prone to break apart when moved from the aircraft collector to the sample mount for analysis. If the microscope needle touches the particle too firmly, then it can easily fall apart into even tinier pieces that are not ideal for analysis, being even harder to handle.

A single IDP can be composed of thousands upon thousands of tiny pieces of silicate rock, each individually just a few hundred nanometres across (a nanometre being one billionth of a metre). These are what give the particle its 'fine-grained' texture. These small pieces of solar nebula dust are mixed together with fragile organic matter, and in space ice will also help to bind the dust and organic material. Once a piece of comet dust makes it to the Earth's surface the ice is obviously long gone, leaving behind a rather holey structure – a dry sponge. The structure of comets can be described as 'unconsolidated': uncompacted and loosely arranged. In contrast, a particle originating from an asteroid is more likely to be consolidated – composed of hard and rocky particles that are compacted together, resulting in a low porosity.

Let's quickly recap the reason for these contrasting structures that are a direct result of the environment within which the objects formed. The fact that the comet-forming region in the outer Solar System is vast, and sparse, means that most comets are separated by at least several hundreds of kilometres. Remember the analogy to those remote farms situated on the outskirts of the city? As a result, comets have been much less likely to experience a collision with another object during their history, allowing them to maintain their original fragile structure – quite simply, a dusty cloud. The sparsely populated comet region is in stark contrast to the relatively congested early inner Solar System, where asteroids and early-forming planetesimals were much more likely to experience collisions with each other. The first few billion years of Solar System formation were a chaotic time, particularly closer to the Sun. This was a time marked by high-energy collisions where impacting objects could either be smashed to pieces or forced together – melted and fused – to produce a well-consolidated larger rock that might have become a planet. As such, the inner belt asteroids are well-consolidated like the planets. The asteroids of the outer belt are not expected to share this history or structure, as they are thought to have formed in the same place as the comets and are much more likely to show many of the same features as the comets, probably even sharing their rather fragile construction.

So, if a fluffy-looking IDP is found in the laboratory, then there's a high chance it contains some primitive solar nebula materials – representing either a comet or an outer belt asteroid – and the particle can be investigated further to discover more about these ancient ingredients. While a standard optical microscope can help the scientists to find the particles at first, such an instrument is stretched to its limits for IDP research at this stage. Scientists need to upgrade to a more advanced and powerful instrument to look at their fluffy dust, something like a scanning electron microscope (SEM).

An SEM scans a focused beam of electrons – negatively charged subatomic particles – across a sample and the electrons in the beam interact with atoms in the sample. The electron-atom interactions produce different signals that allow scientists to determine information about the surface of the sample and its elemental composition. There are a whole range of SEMs that are designed to do slightly different things – studying biological cells, metals or crystals – but the beauty of all of them is that in addition to chemical measurements they can produce high-resolution images of very small objects. This means that when IDPs are analysed, even the tiny individual grains of solar nebula dust that compose the fragile particle can be seen in detail. Remember that these grains can be just a couple of hundred nanometres across. In some cases, depending on the size of the particle and how advanced the microscope is, the elemental composition of individual grains can also be estimated. Using an SEM is often an important step in the laboratory analysis of IDPs, enabling scientists to see the structure of the particle in detail and to pinpoint the location of any interesting grains that warrant further investigation.

Delving into the history of comet dust – an international laboratory effort

There are many, many possibilities for the further analysis of IDPs, depending on what exactly needs to be investigated to answer the scientists' queries. Here I need to take a little step back from the comet dust to show you the bigger picture of everyday scientific investigations. The work a scientist can achieve is usually very much dependent on the budget constraints of the laboratory they work in, as this controls what scientific equipment is available to them. Highly specialised scientific equipment comes with a high price tag, often running into the millions of dollars that stretch already tight laboratory budgets. This means that individual laboratories tend to focus their work around a particular type of instrument, instead of buying all the different instruments available on the market. Hence, different laboratories develop advanced skills, performing a specific type of analysis, meaning that the data they obtain should be of a high quality and trustworthy. When a scientist establishes that further laboratory investigations are required on a sample, for instance to test a new hypothesis, but they don't have the right instruments available in their laboratory to carry them out, they will often aim to collaborate with other scientists. The benefit here is that collaborations bring a range of expertise to the overall study, meaning that more information is gleaned from each special sample. Collaborations are often very important in scientific research, in any field you might wish to name. Without scientists being able to work together, often internationally, they wouldn't be capable of making such impressive progress as they have in most areas.

It's not just about getting raw numbers out of laboratories, though, as the interpretation of any new data is a big, and very important, step in the scientific process. Often data handling and interpretation can be a lengthy process, too, particularly if several laboratories were involved in the overall study. Many different datasets may need to be combined to get the bigger picture and extensive data analysis has to be carried out in order to interpret the story held within the numbers. At some point the leading scientist(s) must agree how to interpret the data so that the results can be reviewed by their peers – other scientists working in similar fields – and published in a scientific journal. Of course, once the work is published it doesn't mean that the conclusion reached by the investigating team is necessarily the correct one. The paper is open to the scientific community to discuss, and dispute if necessary. This often leads to new projects to investigate the exact same thing when the community, or a small group of scientists, disagrees with a published interpretation, or maybe with the way the study was set up. Or they might think they have a better scientific instrument available to them to answer the same question. This is, however, how the scientific process works. It is rare that there is a definitive interpretation from a collection of data. Instead, slowly over time – and often with more experiments – the community works its way towards an accepted consensus. Often scientists won't be able to agree entirely, so the original question remains open to debate. Despite the fact that science always seems to come up with more questions than it answers, I can speak for many scientists when I say that this is exactly part of its appeal and what fuels most of us to keep going.

Probing inside the comets

So, let's get back to the comet dust in the laboratory. Commonly, it is the organic matter and rocky silicates within IDPs that are studied in the most detail, simply because the ice is no longer present in samples once they reach Earth. Although there are many techniques available to enable us to investigate these components, studying the different species of isotopes of various elements they contain is often very informative. Isotope analyses can be very accurate and precise and, as we saw in Chapter 3, they can allow scientists to 'fingerprint' different samples. This often means that scientists can determine the journey the sample has experienced, from the moment it was created to its delivery to Earth. Isotope studies are used to work out where and when the rock formed and in what kind of environment, such as how hot or cold it was.

One of the best instruments to measure a range of isotopes in tiny IDP samples, among others, is that of a secondary ionisation mass spectrometer (SIMS). It is, effectively, an even more specialised type of microscope, but unlike the SEM, which uses an electron beam, a SIMS uses a focused beam of ions, which are negatively or positively charged atoms or molecules. This beam of ions is fired at the sample and when it encounters the surface it literally knocks off some of the particles, causing them to be 'sputtered' away – essentially ejected from the sample. This procedure destroys the surface of the rock, ion by ion. Such destruction is fine when you analyse a large piece of rock, but, for an IDP, sputtering away ions in this manner can demolish the entire rock because, since they are so small, they are not made up of much material to start with. Nevertheless, obliterating samples ion by ion is all done with careful planning and, as the ion beam bombards them, they release their 4.6-billion-year-old secrets in the form of chemical data.

Sputtered ions produced from the sample are focused through a series of lenses and then deflected by a large magnet before entering the mass spectrometer part of the instrument, where they are counted. The magnet, which can weigh more than a tonne, does the job of sorting the ions by their mass – those that are heavier get deflected less by the magnet than those that are lighter. As such, the ions enter the mass spectrometer in slightly different places, dependent on how much they were deflected, and the detectors in the spectrometer detect and count them. The differences in mass between the different isotopes being measured are absolutely minute, just a neutron or two, but these instruments are so sensitive that they can distinguish such tiny differences.

One version of the SIMS instrument, called a NanoSIMS (where 'nano' refers to the size of the focused beam of ions, being just a few hundred nanometres across), can measure very small and precise areas of sample, and can sample several different ion species simultaneously. For example, three of the isotopes of oxygen – oxygen-16, oxygen-17 and oxygen-18 – can all be measured at the same time in the same small region of sample to give an accurate and precise oxygen isotope signature. Sometimes this sampling area can be just a couple of micrometres across – absolutely tiny.

The beauty of the NanoSIMS is that it can also produce a chemical image, or an isotope map, of the dust grain in terms of the ions it's measuring from it. This may sound quite bizarre, but it's a powerful tool for a scientist, as it provides information about where the different isotopes are located within the sample. Thanks to the very small and focused beam of ions, the image of the sample has a very high spatial resolution, meaning it is possible to resolve really small features within it. For example, many IDPs contain presolar grains – those grains from other stars that are older than our own and were caught up in the formation of the Solar System. Presolar grains have an exotic chemical composition – because they weren't formed by our Sun – so they look quite alien in terms of their isotopes. The problem is that they can be exceptionally small – sometimes just tens of nanometres in diameter – which makes them very hard to spot, nigh on impossible with the naked eye, even using a powerful optical microscope. However, when they are analysed in a NanoSIMS to produce an isotope image, they 'pop out' very clearly, as their composition is so different to that of the background rock that was all formed within the Solar System and gives 'normal' isotopic values. When analysing fluffy IDPs composed of hundreds of tiny grains of solar nebula dust, each grain only a few hundred nanometres across, and presolar grains from other stars that happen to be the exact same size, this technique is invaluable to distinguish between the different parts of space that contributed to the sample.

Just like the larger meteorites, it is often the measurement of oxygen isotopes within the rocky silicate solar nebula dust of IDPs that can be the most informative when trying to unearth the history of the sample. Because one of the major elements that forms silicate rock minerals is oxygen, its isotopes are contained in a high enough abundance to measure them accurately in rocky samples. However, when oxygen isotopes are measured in IDP silicates, the amount of oxygen available to measure is severely limited because of their small size. After all, analytical precision is linked directly to counting statistics. The more of a particular atom you can have available to measure (for the spectrometer to count), the more precise the measurement of its isotopes will be, which gives greater confidence in the result. With IDPs, scientists are limited by what they have available, which is a very small piece of dust. Nevertheless, determining the composition of a tiny sample of a comet is better than not measuring a comet at all, even if the same level of precision can't be reached as it could on a larger meteorite.

The life of cometary dust revealed by oxygen

As discussed briefly in Chapter 3, one of the exceptionally useful things about using oxygen to investigate the chemical evolution of the Solar System is that the abundances of its isotopes are affected by how, where and when the rock formed into which they were incorporated. As we learnt, this results in different planetary objects having unique oxygen isotope signatures. For example, a meteorite from Mars has a different oxygen isotope signature to Earth because it formed in a slightly different location in the Solar System.

Therefore, it is interesting to study the oxygen isotope composition of IDPs, as they can inform scientists about a completely different region of the early-forming Solar System – that of the outer disc. When scientists do so, however, they find that the oxygen isotope compositions of the IDPs don't tell a simple story. Their overall composition – the range seen when measurements of all the individual IDPs that have been analysed are considered together – extends to a wider range than, but overlaps with, that measured for all the meteorites. This is an interesting and somewhat unexpected result as it suggests that some of the IDPs – those that are very similar in composition to the meteorites – could have formed within the same location as them. Some IDPs even share the same composition as that which characterises CAIs and the Sun! You'll recall that the Sun, and the solids that formed very early on next to it, are relatively enriched in the lightest of the oxygen isotopes, oxygen-16. Does this mean that some of the comets that these IDPs originated from formed next to the Sun? Probably not. But it is not instantly obvious to scientists why the IDPs should cover such an array of compositions.

If we take a step back we see that, when scientists analyse a collection of IDPs, they are sampling a large selection of different cometary bodies. The range of compositions they measure reflects the fact that the comets that have been sampled formed in a wide variety of locations around the young Sun, and even at slightly different, albeit early, times (say within the first few million years of Solar System history and definitely before the planets were all fully established). As such, each comet is expected to record a somewhat different history, one that is revealed in its oxygen isotope compositions. It's hard to know exactly when and where the comets formed, but it is highly unlikely that their formation was an overnight process that occurred all in one location. Nevertheless, they are not expected to have formed so close to the Sun that they could inherit its distinctive oxygen isotope composition. What we shall say for now is that the comet formation story is not as simple as predicted. Without giving the game away just yet (we'll learn more about this in Chapter 7, as it was the *Stardust* mission that taught scientists to expect the unexpected in comets), it turns out comets don't just contain the earliest solar nebula dust, organics and ice from the outer solar nebula.

It has taken some scientists a leap of faith to accept how these small dust particles might have gained their wide oxygen isotope variability and, in fact, many are still undecided on the reason. Whatever the way, it seems that materials formed within the inner Solar System – those same ingredients that form the asteroids and planets such as CAIs and chondrules – found their way soon after they formed to the infant comet neighbourhood where they were combined with the fine, fluffy, fragile solar nebula dust and fragile organics that existed there. The mechanism for the transporting of inner Solar System materials many AU to the outer disc within the first few million years of history is something that scientists are investigating, and there is, as yet, no consensus. What scientists are finding out is that IDPs really are teaching them that comets are doing their best to avoid adhering to the classic Solar System models. Importantly, they are learning that space missions such as *Stardust* and *Rosetta*, which visited comets up close, have been invaluable in furthering our knowledge of these ancient objects at the same time as providing the essential 'ground truth' to better understand IDPs.

Of course, scientists are not limited to measuring oxygen isotopes to reveal the history of these little particles: there is a whole raft of other isotope systems that are useful for tracing the events that took place in the early Solar System, too. You can probably pick any element in the periodic table and a space geologist will find it, and its isotopes, useful to trace some process or another. Nevertheless, oxygen isotopes are often a good place to start because, as a commonly analysed isotope, the values obtained can be compared directly with the vast amounts of data already in existence for other Solar System objects. It's not just the rocky parts of samples that scientists want to analyse, though, as IDPs also contain a high proportion of organic material. In fact, sometimes organic carbon can constitute up to 80 per cent of a sample (there is virtually no rock in some IDPs). This organic material has an interesting story to tell, too, indicating that some of the organic matter present in these samples might even pre-date the Solar System.

4.6-billion-year-old organic matter, preserved in cometary dust

Commonly, it is isotope systems such as hydrogen, carbon, nitrogen (and even good old oxygen again) that are analysed in the hope of picking apart the history of the carbon molecules that make up the organic matter in IDPs. After all, these are the main elements that compose the organics. In keeping with the fact that the comets are thought to have formed in the cold, outer disc, the hydrogen and nitrogen isotope signatures of their organic matter tend to have compositions indicating they formed in a very cold environment. This is presumed to be either at the very edge of the outer disc or even in cold, distant, interstellar space. In the latter case, this would suggest that some of the organics present in comets were inherited from outside of the Solar System. This is an interesting thought. If the fragile organics formed in interstellar space before being incorporated into the comets, then they must have survived the violent birth of the Solar System. While this might be hard to fathom because the organic matter is very fragile and labile – easily altered – without further work scientists can't say for sure either way.

To help further figure out the cometary histories, the scientists don't look at the organic matter in isolation. While understanding everything about the organic matter itself is useful, seeing how it relates to the rock it is contained within is even better. After all, the silicate rock dust and organic matter ended up being closely associated within the comets – intimately mixed together at a fine scale. While these two components may not share the exact same history, studying them in parallel can potentially shed light on how they ended up together. Understanding whether the organic matter and silicate dust formed in the same place or not is an important first step in figuring out how organic matter forms and survives in space.

To do this, scientists can pick from a whole range of techniques to analyse their samples. Some of these can look at how the elements within the organic matter and silicate rock dust are connected to each other, and others at how the two components are related structurally. Such work has revealed that the organic matter often coats the individual silicate grains, acting like a glue. The way to imagine this is if you take a small piece of rock, something as tiny as a single grain of sand, and roll it in tar, or organic 'goo'. Then if you take many of these tar-ball grains and lump them together you would expect them to all stick to one another. In this way, you can start to imagine the structure of the IDP rock dust and organic matter.

Such findings help to explain why comets, and particles that originate from them, are apparently so fragile. The organic 'glue' holding the rock dust together is not very strong, not really like tar. Instead it's made of loosely formed networks of carbon. The only other component to help hold the structure together is ice, which is rather prone to melting when the comet comes into the inner Solar System. Importantly, however, this close association indicates to scientists that the rocky and organic materials are likely to have formed at a similar time to each other. Perhaps the organic matter formed just after the rock dust grains, such that it was able to wrap around the dust before all of this mixture was captured and incorporated into a larger cometary body. Unravelling the formation and subsequent history of the organic components in IDPs is clearly still an active area of investigation. In less than a century of analysing cometary materials scientists have already learnt a great deal about their history. Future sample return missions to comets would help to accelerate this learning, but at the present time none have yet been funded. So, scientists will continue to analyse the samples they have available and thankfully there are many, even if their total mass is small.

Organic matter isn't just found in comets, of course; it's also very abundant in many asteroids, too. The only problem is that it often can't be analysed in the same way to provide information about how and where it formed. This is because, although the organic matter probably formed at the same time and place as the cometary organic matter, many of the asteroids underwent a complex history of alteration near to the Sun after they formed (as discussed in Chapter 3). As a result, any fragile organics they contained would have been irreparably changed (thermally altered) from the effects of the Sun's heat. The heat will also have acted to melt any ices present in the object, triggering the circulation of liquids throughout the asteroid, causing further (aqueous) alteration as these liquids reacted with different minerals and components present. In such a process, the organic matter is gradually changed in chemical composition and structure. Despite this, while the altered organics aren't of as much help for understanding where they might have originated, they still have their uses because they can trace a slightly later history, the one experienced by the asteroid after it formed. In fact, organic matter is often more sensitive to changes occurring in the rock than the rocky components themselves, meaning that the thermal and aqueous changes experienced by the organic matter are easily traceable by scientific analysis. This in turn can help scientists trace where the asteroid went (how close to the Sun), and what else it experienced before it settled down in the asteroid belt.

As we've seen, comets and the samples that originate from them, such as IDPs, are essential for probing into the origins of our Solar System. They can also help scientists to learn about the organic matter that may very well account for our presence as humans in this corner of the galaxy. Sample return missions such as *Stardust* and other missions that have visited comets up close, most notably *Rosetta*, have helped to accelerate our understanding of comets, but IDPs still play an important role. Luckily, with careful collection and analysis on Earth, small cometary IDP samples are surprisingly common and often well-preserved, despite their long journey to Earth. It is simply their small size that makes them challenging to analyse, requiring highly specialised equipment and scientists with many years of experience. However, these precious samples currently give scientists their best chance of understanding our place in the Solar System, without the expense and technical challenges of collecting comet samples in space and returning them to Earth. Despite this, obtaining samples directly from a cometary nucleus would still be incredibly useful, especially if the cometary ices could also be returned to Earth intact, something that is not possible for IDPs that come rocketing through the atmosphere. While we must wait patiently for such a mission to come to fruition, it will be the little IDPs that continue to serve an important role as messengers from 4.6 billion years ago. We can only hope that with further analysis they will reveal some more of their long-held secrets to allow scientists to piece together what really happened all those years ago.

CHAPTER 5

Water and Life on Earth and in Space

When I was younger, I loved the idea that there could be some extraterrestrial organisms out there that had far surpassed our intelligence and knowledge, and had figured out how to live among us on Earth, undetected. My wild imagination was probably spurred on by my reading and watching of too much science fiction, *The X-Files* being a personal favourite. However, I'm clearly not the only one who has been captivated by this subject. To be fair, most people probably don't think there are alien creatures living inconspicuously around them, but the idea that there is life out there on another planet in the Solar System, or somewhere else in the Universe, is a popular one. Of course, just like me, many people are led by science-fiction narratives, but the search for life in the Solar System, and elsewhere, has been a key research topic internationally for many decades. The questions seem simple: are we alone in the Universe and what might life on other planets look like if it were to exist? However, answering these questions is not necessarily straightforward. Unless my childhood ideas were correct, life in our Solar System, should it exist, is certainly not very advanced or we would have seen it by now. Part of the reason we, as humans, are so interested in whether organisms exist elsewhere is that we still don't know how life on our planet got here, where it came from, or how and why it took hold. If we really are alone in the Universe, then it would be good to understand why our planet is apparently so special, sitting in an apparently barren Solar System, possibly even a barren Universe.

Added to this is the fact that scientists don't fully understand where Earth's water came from, and this is important as it turns out that it's an essential constituent for life. Indeed, understanding where water came from has important implications for the emergence of life not just here on Earth but anywhere else we look, too. Certainly, Earth appears to be the only planet in the Solar System that holds liquid water at its surface, and it is also the only place we know that hosts life (as we know it). The chances that these two facts are related is high and the presence of water on Earth appears to have been essential for life to have arisen. Water is a very important solvent wherever you are in the Solar System, not just on Earth, particularly if you want to embrace biological processes. Without it, quite simply, we wouldn't be here and neither would any of the life that has existed on this planet throughout history. But what about life on other planets? Well, it hasn't made itself obvious yet, but that doesn't mean it isn't there, or hasn't existed in the past. But the big question is whether life on other planets, should it exist, also need water to survive? Some seemingly simple, yet big, questions remain.

Where is all the water in the Solar System?

When it comes to water in the rest of the Solar System, despite first appearances, there is actually plenty of it, it's not just confined to Earth. In fact, water is abundant from the comets of the cold Kuiper Belt all the way to the constantly shadowed polar craters of Mercury, the closest planet to the Sun. The problem is, these places host water as ice, and life has trouble existing and reproducing in solid water, or in a solid version of any solvent, in fact. Even a gas isn't very helpful for life, as it can't maintain a stable enough environment for the required chemical reactions. In general, liquid solvents are required to enable chemical reactions and they are also effective at physically transporting materials. Water is known as a 'universal solvent' and it's a polar molecule, meaning it has positively and negatively charged ends – more positive near the hydrogen and more negative near the oxygen. This permits it to dissolve many chemicals, allowing them to recombine in different configurations – an important attribute for the many organic compounds. Although other solvents might be capable of supporting life, many are not stable at the range of temperatures that we think lifeforms can cope with, even those adapted to more extreme environments like the bugs that survive on the exceedingly hot black smokers miles under the oceans on Earth. But that's not to say that there aren't some organisms out there capable of surviving in conditions that we don't think are suitable for life as we know it. We should definitely keep an open mind. However, scientists think it is most likely that life in the Solar System would prefer to find its feet in water. After all, water is liquid at a range of biologically useful temperatures and it's composed of elements that are very abundant in the Universe – hydrogen and oxygen. Because of its abundance and value in supporting chemical reactions, statistically speaking, water seems to be the most likely solvent to be capable of supporting life. So, at least initially, scientists looking for life elsewhere in the Solar System search for liquid water first.

Luckily, liquid water seems to exist, and to have existed, elsewhere in the Solar System, just not currently at a planetary surface. Hence, space exploration missions to investigate such places are required before scientists can make conclusions about whether they host life, or have hosted it in the past. A promising location to search for life in the Solar System is Europa, one of the moons of Jupiter. Europa is known to have a core of iron, a rocky mantle and an ocean of salty liquid water below a surface of ice; it therefore, sounds a bit Earth-like in some ways. Europa will most likely be explored with spacecraft in the future – in fact, NASA's *Europa Clipper* mission, scheduled for launch in the 2020s, is planned to investigate its oceans. We'll just have to wait a while to see if any life is lurking there.

Life on Earth

Even though the existence of life elsewhere in our beautiful and diverse Solar System has not yet been proven, the fact that water is present in liquid form on Earth means that life has been able to take hold, and thrive, here. As we've discussed, whether water is needed elsewhere in space for life to have similarly taken hold is unknown. Nevertheless, understanding how life managed to achieve such a stronghold on Earth might help scientists to investigate the potential for life on other planets within, and even outside of, the Solar System. Some of the earliest pieces of evidence for life on Earth are around 3.7 billion years old, found in ancient rocks from Greenland. The fossilised microbial mats found in these rocks, known as stromatolites, are formed of bio-layers of cyanobacteria. They were the dominant life form for over 2 billion years of Earth history and were responsible for creating Earth's atmospheric oxygen. The rise of oxygen in our atmosphere was an important phase of Earth's evolution, as it fuelled the Cambrian explosion, a time when many major groups of organisms appeared in a short timescale, over about 40 million years. It's also a time when aerobic forms of life started to evolve. As such, we have a lot to thank this long-lived species for. Such microbial mats are still around on Earth today, although they tend to prefer places where other organisms struggle to survive, favouring hypersaline, highly alkaline, low nutrient and high or low temperature regions. Today they have been found in Australia, the Bahamas, the Indian Ocean and even Yellowstone National Park. It's quite amazing to think about the amount of Earth history these cyanobacteria species have witnessed, surviving quite a few mass extinctions that killed off most other life on Earth.

Stromatolites may be biologically simple compared with much of the other complex life that has arisen on Earth since. However, despite their biotic naivety, the question still remains how they came into being at all? Whatever these cells evolved from was presumably even simpler. If something preceded them, then it's possible that life began on Earth as early as 3.8 billion years ago, or even earlier. This happens to be very close in time to the oldest rocks that exist on Earth today. Although life could exist without rocks, the fact that Earth was a hot, molten ball of rock for at least the first few hundred million years of its existence suggests that it would've had a hard time surviving during this period. Added to this is the fact that Earth was experiencing frequent major impacts from space around 3.8 to 4 billion years ago, making it an uninviting and unstable environment for life to try and find a foothold. These cosmic impacts would have constantly caused Earth to experience catastrophic changes to its environment, even vaporising early oceans that might have formed. But it is perhaps intriguing that the end of this violent phase of Solar System history coincides with the earliest evidence for life on our planet. We should, perhaps, question whether the two are related. Maybe as the Earth entered a calmer phase, with fewer violent impacts onto its surface, life was then safe to explore. In addition, it is unclear what kind of an atmosphere Earth had very early on. It certainly didn't appear to contain much oxygen for quite a few billion years, but it probably had carbon and methane. If any life evolved very early on, then it would have had to survive anaerobically.

One of the reasons scientists can't know for sure if life began a lot earlier than around 3.8 billion years ago is that fossilised evidence, if it existed, has not necessarily been preserved, due to the action of plate tectonics. Another possibility is that perhaps life emerged and became extinct more than once on Earth. Although there is no evidence for such a scenario, if it were found to be true then it would make it seem easy for life to begin, since it began more than once. If that was the case, then we might expect life to be found more frequently throughout the galaxy. At the moment, we appear to be alone, so scientists prefer the idea that life emerged only once on Earth.

Anyway, it's all well and good knowing roughly, even if not exactly, when life took hold on Earth, but it still doesn't help us to know the reason *why* life appeared. Had life been sitting around dormant as part of the geology since Earth's inception or was it delivered onto the surface later once things had calmed down a bit? When I talk about 'life' being delivered to Earth I don't actually mean life itself (*i.e.* living cells) but instead the ingredients for life (*i.e.* carbon, hydrogen, oxygen, nitrogen and their compounds). Some scientists like the idea of panspermia, of which the literal translation is 'seeds everywhere', which deals with the theory that life exists throughout the Universe. Panspermia says that microorganisms can survive the harsh conditions in space and that they can be transported between planets and star systems. Since scientists haven't observed living things elsewhere in the Universe – not even a very basic microorganism – this theory is currently unproven. Instead, scientists have detected carbon and its associated elements in many places – pretty much everywhere they've looked, in fact. So, although it can't be said for certain, it currently seems more likely that 'life' itself was not delivered to Earth from outer space, but rather the ingredients for life. These basic constituents may have also been delivered to other planets, but it was on Earth that they are known to have encountered a hospitable environment in which life could evolve. Whether other planets also provided a comfortable welcome is yet to be discovered.

Earth's water

The same questions we ask about life on Earth can equally apply to water: was it here from the very beginning or was it delivered later on from outer space? Since scientists are pretty sure water was needed for life to commence and continue evolving, focusing on water allows them to understand more about life itself. Obviously, Earth at the present day contains a lot of water, and a variety of interesting life forms, so let's work our way back to the beginning to see if we can figure out where it all came from.

The original water of the Solar System was inherited directly from ices in interstellar space that were swept up in the early solar nebula as it formed. The problem is that scientists don't know if these ancient interstellar water ices could have survived the harsh conditions of the birth of the Sun and the resultant formation of the planets. If it can be proven that ancient interstellar water survived these processes and is, to this day, incorporated into all of the Solar System objects that formed from the solar nebula – including the planets, comets and asteroids – then it could have really important implications for other star systems and the hunt for life in the Universe. If our Solar System's formation is typical, then there's no reason why the same couldn't have happened in other planetary systems, making the search for life elsewhere potentially more fruitful. However, it seems likely that the ancient interstellar ices could have been destroyed during the birth of the Sun, so that the Solar System's water today is 'solar': the ancient ice was modified in chemical reactions brought about by the formation of the Sun. Such a scenario would suggest that the amount of water seen, if any, in a star system is dependent on the individual star, making the chances of life elsewhere slightly less likely.

Regardless of which of these scenarios is correct, there is another issue. As we've seen, the infant Earth is not thought to have been a particularly obliging environment for collecting and saving water, and it certainly isn't thought to have been hospitable to life. Early on, Earth's surface temperature was very high, possibly ranging anywhere from 700 to 1,200°C – temperatures that mean the Earth was not necessarily solid. These very high temperatures would have been partly maintained by the seemingly never-ending series of high-speed cosmic impacts that marked the early days of Solar System formation. Such impacts would have helped to sustain Earth's molten state, even down to deep levels within the planet, for a considerable amount of time. Earth's surface would have resembled a magma – literally molten rock. Any water contained in the magma – being one of a host of volatile molecules, all of which have relatively low boiling points – would easily have evaporated away to space.

As such, it seems intuitive to suggest that Earth would have struggled to retain its early inherited water to the present day, regardless of whether that water was interstellar or solar in origin. Because of this, many scientists agree with the idea that Earth lost its early water, or at least much of it, through evaporation before accepting a new delivery from outer space, once it had cooled down a little bit. The most plausible agents to deliver Earth's new inventory of water, and other volatiles, are the comets and asteroids. Whether large or small, many comets and asteroids are known to contain abundant water ice and other volatiles. In fact, even the carbonaceous chondrite meteorites are around 10 per cent water by mass and some objects in the asteroid belt easily contain more water than this. For example, it is calculated that the dwarf planet of Ceres could contain more fresh water than Earth. A few of these water-rich objects colliding with Earth could easily account for Earth's water, even if they were much smaller than Ceres.

The result of impacts

The pockmarked surface of the Moon provides evidence to suggest that a range of objects impacted the early Earth-Moon system, and the other rocky planets, plenty of times in the past, allowing ample opportunity for water to have been parachuted in from space. However, while it is known that comets and asteroids bombarded the early Earth, and that they contain volatiles, their impacts could have instead helped to de-gas the planet at the same time. I'll liken this process to a stage of baking a cake. Let's think of the Earth as a lump of cake batter. When you put the batter in the cake tin prior to baking, you can tap the tin on the counter top and all the gas bubbles work their way to the top of the batter and pop out. I admit, it's not exactly the same, but a big impact on Earth would have shocked it into releasing its trapped volatiles such as water.

However, it wasn't just really massive comets and asteroids that collided with Earth, but a whole range of different-sized objects, some of which were dust-sized, such as the interplanetary dust particles discussed in Chapter 4. These smaller cometary and asteroidal objects wouldn't have impacted the Earth with any noticeable force: they are too small. Instead, they would have settled onto the surface imperceptibly, being almost certain to deliver with them their, albeit small, inventories of volatile space goodies, and even organic material. This period of Earth history was obviously a delicate balance between less frequent large impacts causing Earth's volatiles to de-gas, and more frequent small but gentle 'impacts' (or 'arrivals' of objects) that were able to deliver volatiles. If smaller objects were constantly raining down to Earth, as they are thought to have been, then they could easily represent a significant delivery source of water and organics.

If Earth was collecting up its water from incoming comets and asteroids, the timing of their impacts might have also been important. If the impact of water-rich objects preceded a really large impact, such as that which formed the Moon, then any water the Earth had accumulated might have been lost, being de-gassed away into space (like the bubbles leaving the cake batter). However, it is possible that Earth could have sequestered some of its early inherited volatiles away to great depths in its interior, protecting them from subsequent de-gassing brought about by impacts occurring at the surface. As you can see, the possibilities seem to be endless, which goes part of the way to explain why scientists have a hard time picking apart the story of water and life on Earth.

Another reason why it's hard to say for definite where Earth's water came from, and how much there was at any point in history, is that it's very challenging to estimate how much water the Earth contains at the present day. Even the best estimates produce unsurprisingly unimaginably huge amounts of water: around 1.4 billion km^3. It's not just the water in our oceans, rivers and atmosphere that needs to be accounted for, there's also the water that's dissolved in the interior of the Earth, as part of the rocks that form the mantle (and even the core), which is even trickier to estimate. By taking conservative estimates for the amount of water thought to be contained inside the Earth at the present day, and lower estimates for the water brought in by objects that have collided with it – whether they were comets or asteroids – scientists can calculate how much water was delivered during the later stages of the Late Heavy Bombardment (LHB). The estimates suggest that the delivery of water during this time – around 3.8 billion years ago – could account for at least some, potentially 10 per cent, if not all the water Earth has today. While it is still a possibility that some of Earth's hydrosphere water came from de-gassing of the planet early on, it is a very real possibility that the rest was delivered from outer space.

Extraterrestrial organic material

As hinted already, it's not just water that the comets and asteroids brought with them, as the impacts almost certainly delivered organic material at the same time. Some estimates suggest that around 10^{16} to 10^{18} kg of organic compounds were delivered in the later stages of the LHB. This value happens to be a few orders of magnitude larger than the organic carbon estimated to be present in the Earth's biosphere today. So, at least in this respect, it seems there is serious potential for Earth's organic compounds to have come from outer space.

This means that even if Earth didn't care for its early-accumulated organic material very well, then it easily could have been replaced later. Phew, that's a relief, because those high temperatures early in Earth history certainly weren't very helpful for preserving all that organic matter, not to mention the effects of the huge shock pressures related to large impacts, too. Organic compounds are, after all, fragile and sensitive. Another key factor that suggests life, or the precursors for it, might have come hurtling to Earth on the back of a comet (or asteroid) is that, as mentioned previously, scientists think the early Earth contained little to no oxygen. This would have given the planet a chemically reducing atmosphere, composed mainly of carbon dioxide, methane and nitrogen. Such an environment is not conducive to forming organic matter from carbon. At best, life could only have formed rather inefficiently, if at all. The supply of organic matter from an extraterrestrial source could, therefore, have played an important role in the initiation of life on Earth, bringing in just enough organic matter at just the right time when there was sufficient oxygen.

You may be mistaken in thinking that organic matter is unique to Earth, but it is everywhere throughout the Universe. The atoms that compose living beings – carbon, hydrogen, oxygen, nitrogen, phosphorus and sulphur – are all made in stars. If we focus on carbon, since it is the most important for life on Earth, it is produced during fusion reactions in stars. However, towards the end of a star's lifetime it isn't particularly good at holding onto its atoms and they tend to be forcefully ejected into interstellar space. Where interstellar space becomes denser – because more atoms have collected together in clouds – it is possible to form simple, as well as complex, carbon molecules from all of this ejected star carbon. The materials for life – prebiotic organics – were literally sitting out there even before our Solar System, or any planetary system, started to form. When our Solar System came into being, forming from the collapse of an interstellar cloud that became the early solar nebula, these carbon compounds were incorporated into the planets, comets and asteroids. The chaotic and dynamic environment of the early solar nebula, involving sometimes exceptionally high temperatures, radiation, shock waves and even lightning, meant that some of these carbon molecules were highly processed to produce new compounds. The result is that our Solar System's inventory of organic matter was a mixture of pristine interstellar and highly processed material.

Organic matter doesn't necessarily mean life

It would be useful here to briefly expand upon this key Solar System ingredient – organic matter. When we refer to organic matter in space rocks it can often be confused with life itself and in the first instance it definitely shouldn't be. However, carbon is a key element in the evolution of prebiotic material. Understanding the journey of organic molecules from molecular clouds to complex life forms on Earth, and possibly other planets, is of obvious importance if we want to comprehend where we came from. When we refer to organic matter in this book we are mostly referring to the element carbon and molecules made from it. There is nothing necessarily biological about this material. The simplest of the organic molecules is methane, a single carbon atom bonded to four hydrogen atoms. Methane has been detected in comets, asteroids and other planets, including Mars. Next up in molecule size and complexity is ethane, which is simply two carbon atoms bonded to each other, with each of those bonded to three hydrogen atoms. These simple molecules may sound a bit dull, and they aren't exactly able to give us dinosaurs straight away, but they are, nonetheless, important for realising where life came from.

The really interesting and useful thing about carbon is that it easily bonds with other carbon atoms, meaning that it can form vast carbon chains that act as a backbone for other atoms to join. When those other atoms happen to be elements such as hydrogen, nitrogen and oxygen, it can allow the formation of amino acids, fats, lipids and, most excitingly, the nucleobases that form DNA (deoxyribonucleic acid) and RNA (ribonucleic acid) – the basic building blocks for life. If we understood how the large-scale production of these molecules came about on the early Earth, then we'd be quite close to grasping how life itself started.

What you may not realise is that the formation of amino acids is not something that is restricted to Earth. In fact, more than 70 extraterrestrial amino acids have been identified, in addition to many other organic compounds, in a single meteorite known as Murchison, a rock of asteroidal origin. Yes, that's right, amino acids exist in space. Amino acids are used to build proteins, which are used by life to make biological structures, such as muscle and cartilage, and to speed up or regulate chemical reactions. They are hugely important for life. These compounds are only found in trace amounts in meteorites, and were only discovered through careful laboratory analysis of rock samples on Earth. However, they have also been detected in lunar rock samples, so, although as yet unproven, it certainly seems possible that amino acids are present on other planets. After all, it was not just Earth that was bombarded by comets and asteroids around

4 billion years ago. The problem is that if amino acids are present on other planets, but also in trace amounts, then they are currently below detection limits for spacecraft instruments, so scientists will probably need to return rock samples from the surfaces of these other planets to Earth to investigate this matter further.

Why is Earth apparently so special?

Since all the inner rocky planets should have had similar inventories of water and organic material delivered to them during impacts in history, the big question still remains as to why Earth so obviously hosts life when the other planets apparently do not. It might just be that Earth is in exactly the right position for life to take hold. We are sometimes referred to as being within the 'Goldilocks zone'; it's not too hot or too cold, being the perfect distance away from the Sun that liquid water can be a constant feature. Therefore, astronomers interested in the search for life elsewhere in the Universe focus their efforts on planets orbiting other stars at a 'comfortable' distance, such that liquid water is a possibility. It is these planets that might have a chance of hosting life. The nearest of these so-called Goldilocks exoplanets is 14 light years away, though, so it isn't somewhere we'll be visiting with a spacecraft anytime soon. Nevertheless, until life is found elsewhere in the Universe, the reason why it is so abundant on Earth might remain a mystery

It might be that planet Earth is unique, beginning its existence with all of the necessary ingredients for life in place, which it managed to hold on to for all of its long and remarkable history. It is also a possibility that the asteroids and comets that collided with the Earth might have been destroyed, literally vaporised, during attempted delivery of their volatiles and organics because of high impact temperatures and pressures related to extreme collisional velocities. After all, amino acids are relatively unstable in water, so even if they were delivered to Earth during collisions in the past, there's a high chance they broke down to simpler compounds once on the surface. But such simpler compounds are still useful and could be the starting point for abiotic synthesis that could eventually build up to create the first forms of life. Despite all of this, it might be that Earth didn't need all of the organic matter in comets and asteroids to survive impact for life to take hold: maybe it just needed a very small amount of it to survive.

Measuring the Solar System's water

An obvious place for scientists to start investigating whether the Earth received deliveries of water and organic material from outer space is to measure the composition of the water and organic material on Earth today and see if it resembles that in the small space objects that share our Solar System. The lucky thing is, water isn't just water, it comes in lots of different forms, and so too does organic matter. So it is possible to test whether they are the same, or not, in different space objects.

Sure enough, we know we can get salty and fresh water – a difference that is easily detectable with a simple test using a human tongue. However, there is another type of water that our taste buds won't be able to distinguish. Despite water only containing three atoms in its molecule – one being oxygen and the other two being hydrogen – these elements can exist in several different forms (as isotopes of each other) but still make water. If we focus on the hydrogen, one of its isotopes is deuterium (with chemical notation ^{2}H or D), which is very slightly heavier than standard hydrogen (with chemical notation ^{1}H) because it contains a neutron in its atom, rather than just a single proton. In so-called 'heavy water' the lighter hydrogen atoms are replaced partly, or wholly, with deuterium atom(s) and the very slight difference in masses between the hydrogen and deuterium isotopes results in subtle differences in their behaviour during chemical and physical reactions. As a result, the ratio of deuterium to hydrogen (giving D/H ratio) in water molecules can tell scientists about the conditions under which the molecules formed.

This may sound complicated but it is, in fact, quite simple. Just think of D/H as a proxy for the type of environment where the water formed; whether the water the scientists are studying formed in a very cold or hot place, such as interstellar space or near the Sun, respectively. In this way, and like many other isotope systems that scientists can measure in rocks to investigate how they formed, and what they are related to, the D/H ratio of water in a rock can be used as a sort of fingerprint.

Let's get stuck in to some more of the science now, but stay with me because it'll help us to understand where our water came from, whether we're drinking water from the Sun or from interstellar space. From theoretical studies, the water in interstellar space is predicted to be enriched in deuterium, reflecting its formation in a very cold environment, at temperatures less than 50 Kelvin (-200°C). So, it has a high D/H ratio. Before the formation of the Sun, water in the molecular cloud would have had a similarly high D/H, being inherited directly from interstellar space. When the solar nebula came together and formed the Sun, the D-rich water present in the inner parts of the hot protoplanetary disc would have undergone rapid isotopic exchange reactions with other hydrogen-bearing species (such as H_2 gas) that acted to dramatically lower its D/H.

Water in the outer disc – where it was much colder – underwent more sluggish exchange reactions that allowed it to maintain its original and high D/H, at least initially. Such isotopic exchange chemical reactions, being dependent on temperature, meant that the D/H ratio of the early protoplanetary disc increased with distance from the Sun. As such, it should follow that the rocky inner planets, and possibly the asteroids that formed in the same place close to the Sun, are expected to have incorporated 'solar' water with low D/H, thanks to the isotopic exchange chemical reactions that took place in this region. However, comets formed on the edge of the solar nebula, far from the Sun, are expected to have mopped up D-rich interstellar ices, inheriting their cold interstellar signature.

Thanks to the combined efforts of cosmochemists and astronomers, the composition of water in interstellar space, asteroids, comets and planets can be measured in meteorites in the laboratory, and/or remotely by telescope. These measurements provide information about the true range of D/H values throughout our Solar System, and beyond. These values can be used to test the predictions made by scientists and, in turn, help them to figure out where Earth's water came from.

Fortunately, measurements of Solar System objects fall in general agreement with the predictions of the theoretical model described above. Water compositions measured for bodies in the inner Solar System, including on Mars, the Moon, Earth and Vesta (the largest object in the asteroid belt), are dominated by the lighter isotope of hydrogen (^1H), whereas water compositions in the outer Solar System, as measured in various comets and Enceladus (an icy moon of Saturn), are generally heavier, being more deuterium (^2H)-rich.

However, unfortunately this is not quite the whole story and there are two problems. Firstly, the water measured in inner Solar System objects, despite its low D/H, is still too rich in deuterium to have been formed by interstellar ices that were completely reprocessed by isotopic chemical reactions near the Sun (*i.e.* it should have even lower D/H). Secondly, when scientists look in detail at the many measurements of the composition of water in different comets and asteroids, they find that they don't all tell the expected story. We'll come back to the latter issue, but first let's focus on the problem of there being apparently too much deuterium in the inner Solar System.

The heavy water problem

When scientists looked at modelling the evolution of water in the Solar System they began with a solar nebula that contained no deuterium (*i.e.* an early Solar System that contained no interstellar ices because they had all been destroyed by isotopic exchange during the formation of the Sun). However, what the scientists found was very interesting. They couldn't reproduce the D/H ratios that had been measured in the Solar System objects so far, whether they be those of the Earth, Mars, Vesta, the asteroids or even some of the comets. The water compositions in all of these objects were more deuterium-rich than would be expected if the Sun was responsible for the production of all their water. The D/H ratios produced by the scientists' models were just too low.

The way the scientists accounted for the observed higher D/H ratios is that the Solar System objects *must* have incorporated some interstellar ice (with high D/H) during their formation. This suggestion is important because it means that some ancient interstellar ice must have survived the tumultuous and chaotic formation of the Solar System. When scientists made computer models of such a scenario, they found that as little as 7 per cent but as high as 50 per cent of the water found on Earth could be very ancient (having formed in cold interstellar space). It's rather fun – and perhaps slightly off-putting – to think we might be drinking water that is older than the Solar System. For comets, the scientists' simulations found that it could be from as little as 14 per cent but as much as 60 to 100 per cent of their water being inherited directly from interstellar space. This isn't a surprise since we know that most of the comets formed very far from the Sun and so we could expect them to be composed of at least some, if not all, interstellar water, as they were very close to interstellar space.

The other thing is that we also know that the comets formed over a range of heliocentric distances, meaning that they formed at a range of, albeit remote, distances from the Sun. For this reason, they can probably be expected to show some natural variation in D/H ratio depending on the exact location where they formed (if D/H ratio truly correlates with distance from the Sun, that is). The comets also formed over a period of time. This is an unknown quantity but is probably at least a few million years. Some comets will have formed exceptionally early, straight after the birth of the solar nebula, presumably inheriting the highest D/H (most interstellar-type signature) before it had been processed away by the Sun. Other comets will have formed a little later, but still before the birth of the planets. These ones could easily have inherited some of the reprocessed, lower D/H water that was eventually transported around the solar nebula from near the Sun where it formed. The D/H of individual objects, therefore, could easily be related to exactly where, and when, they formed. Hence, studying the D/H of different space objects can potentially be very informative.

While such models are useful in helping to account for the measurements currently available of D/H in the Solar System, this is science that is very much still in progress. Many more objects will need to be analysed to build up a strictly accurate picture of water in the Solar System. In particular, measuring the D/H ratio of more of the icy satellites of Jupiter and Saturn would help scientists to test whether hydrogen isotope compositions (D/H) are truly correlated with distance from the Sun.

Water in comets and asteroids

This brings us back to the comets and asteroids, and the fact that they don't all adhere neatly to the same story. Sure enough, the D/H picture painted by the measurement of comets is intricate. Even though the vast majority of comets now reside in the outer Solar System, they were almost all formed closer to the Sun than the positions where they now reside, and were subsequently thrown further out. Despite this, the comets still formed at the largest heliocentric distances of any Solar System objects. You'll recall that the Kuiper Belt comets are thought to have formed further out than the Oort Cloud comets, even though it is the Oort Cloud comets that are now the furthest away. This causes some confusion as to what can be expected for the D/H ratios of these different objects.

Early observations of a Kuiper Belt comet, 103P/Hartley 2, by ESA's *Herschel Space Observatory* found its D/H ratio to

Above: The nearly-full Moon (99.9 per cent phase) taken May, 2017. Craters can be seen across the lunar surface, and are particularly obvious on the shadowed left side of the image.

Left: Atacama Large Milimeter Array (ALMA) image of a protoplanetary disc surrounding a young star. Newly forming planets about the size of Saturn can be seen in orbit, revealed by the imprint left in the gas and dust.

Left: A meteorite on blue ice in Antarctica found by an ANSMET (Antartic Search for Meteorites) team.

Left: ER-2 high-altitude aircraft in flight over the southern Sierra Nevada.

Below: Clean laboratories in Building 31, *Stardust* laboratory at the Johnson Space Centre, NASA.

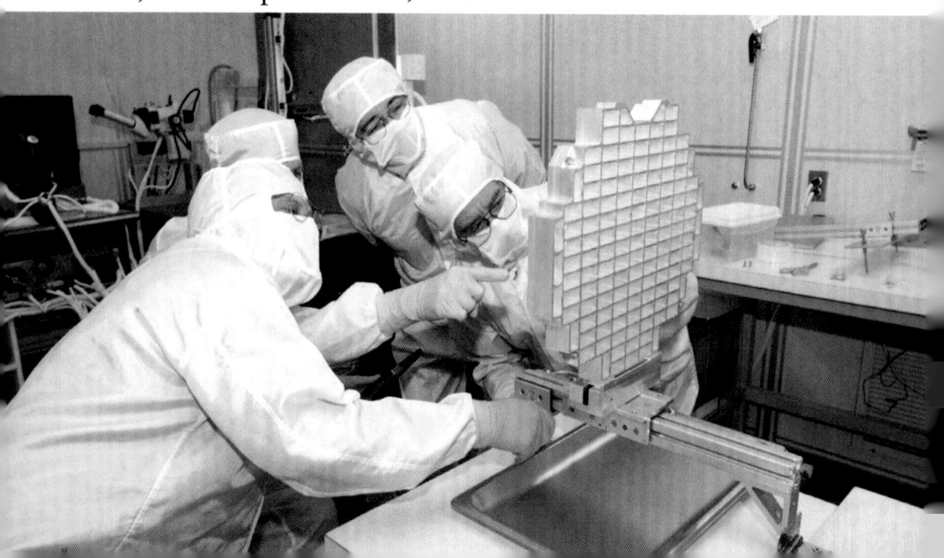

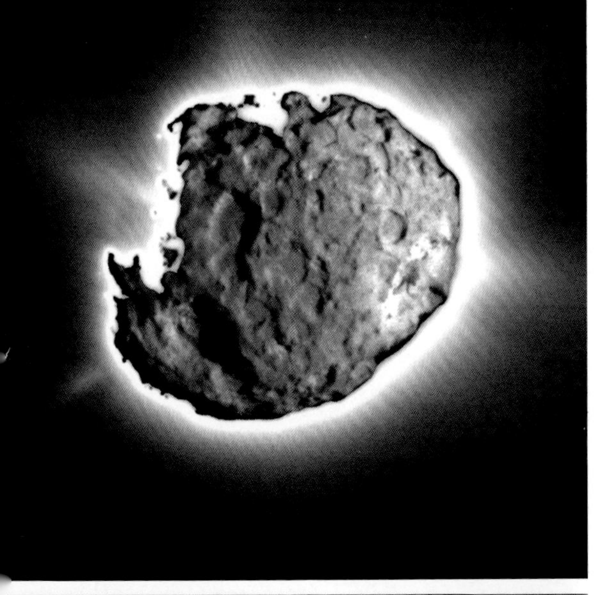

Left: Composite image of Comet Wild2 taken by *Stardust*'s navigation camera during close approach on 2 January 2004. Wild2's active surface, dust jets and gas streams can be seen. The comet is about 5km in diameter.

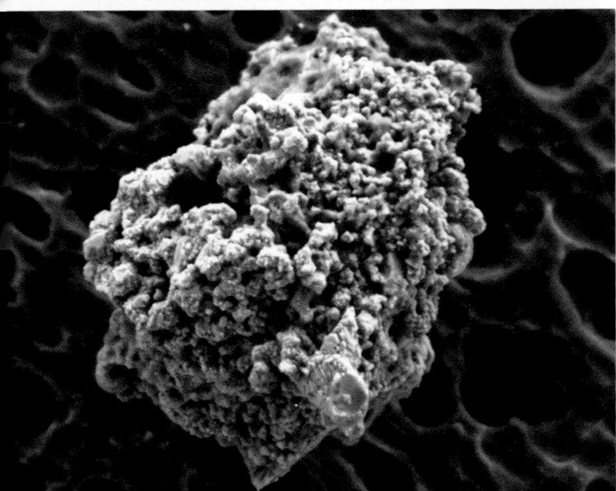

Left: Interplanetary dust particle magnified by Scanning Electron Microscope imaging. It is approximately 7 micrometres across.

Left: *Philae*'s Ptolemy instrument in a clean room prior to flight.

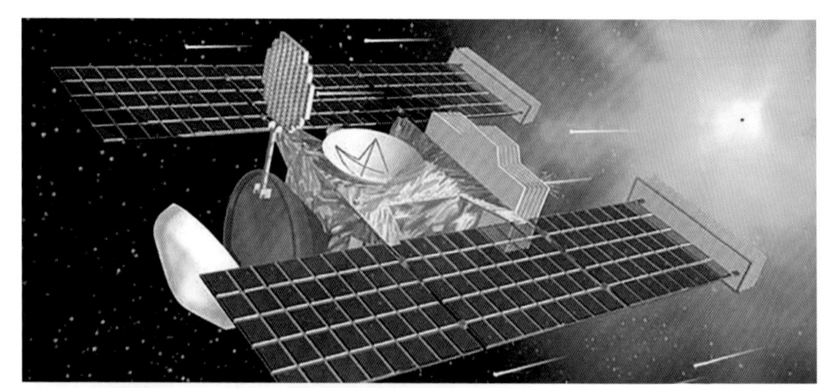

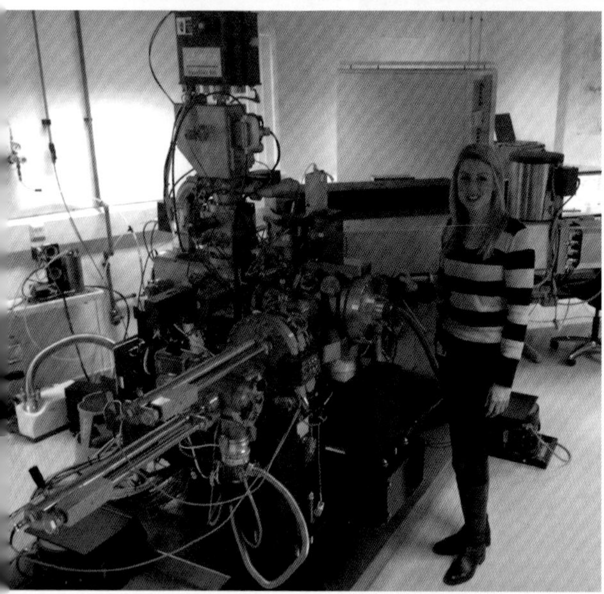

Above: Artist's impression of *Stardust* spacecraft with its 'tennis-racquet' style collector.

Left: The Open University NanoSIMS 50L instrument, with the author.

Below: *Stardust* sample return capsule after touchdown in the desert on 15 January 2006.

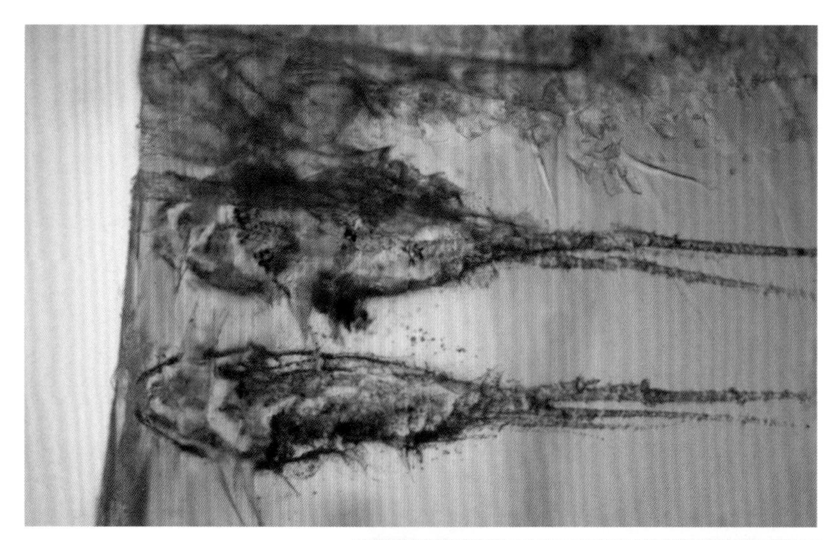

Above: Two burrows left by Wild2 comet particles after they impacted the *Stardust* aerogel collector. Each track is approximately 2mm long.

Right: Viewed in a clean lab in building 31, Johnson Space Center, NASA. *Stardust* aerogel collector tray with some blocks removed for analysis.

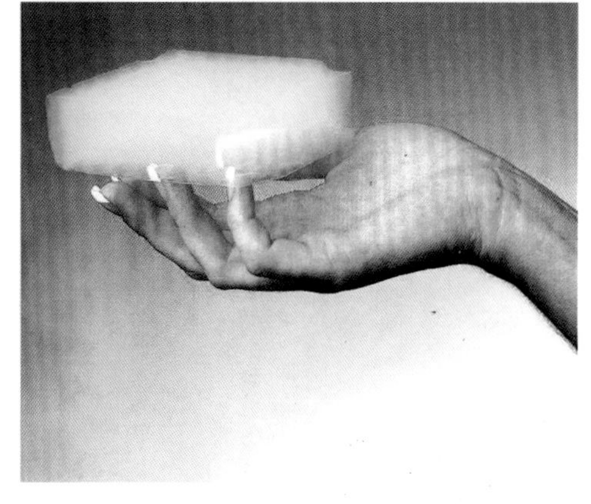

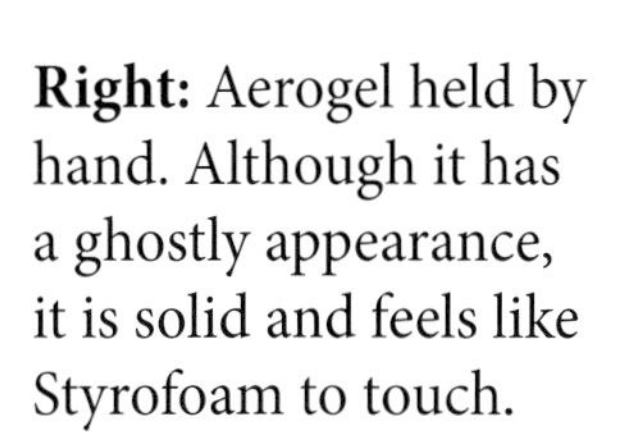

Right: Aerogel held by hand. Although it has a ghostly appearance, it is solid and feels like Styrofoam to touch.

Left: Artist's impression of the deployment of *Rosetta*'s *Philae* lander from the orbiter to comet 67P/C-G.

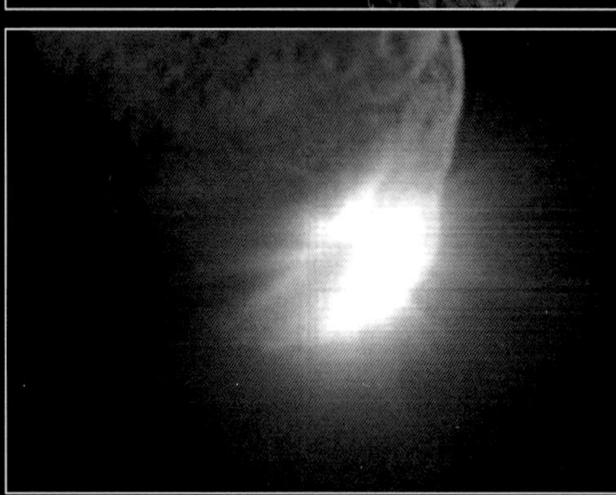

Left: Initial ejecta from Tempel 1 resulting from the NASA *Deep Impact* bolide collision on 3 July 2015.

Below: *Rosetta*'s *Philae* lander spotted on the surface of 67P/C-G almost two years after touchdown. *Philae*'s three legs are seen extending from its metre-wide body.

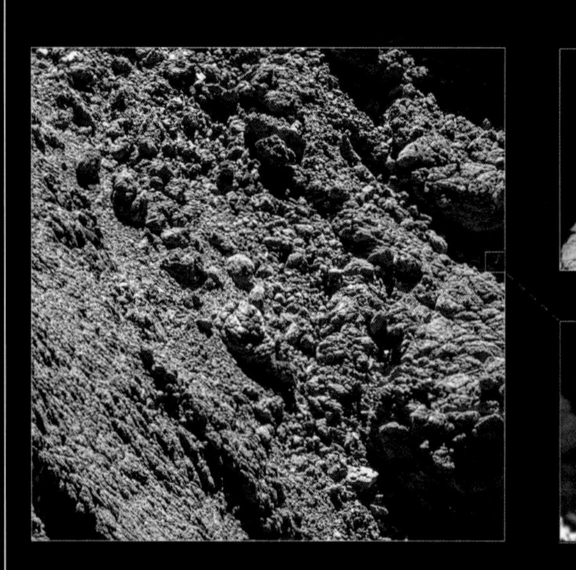

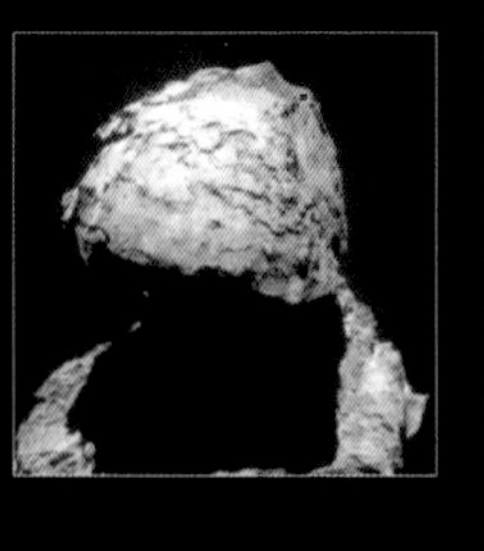

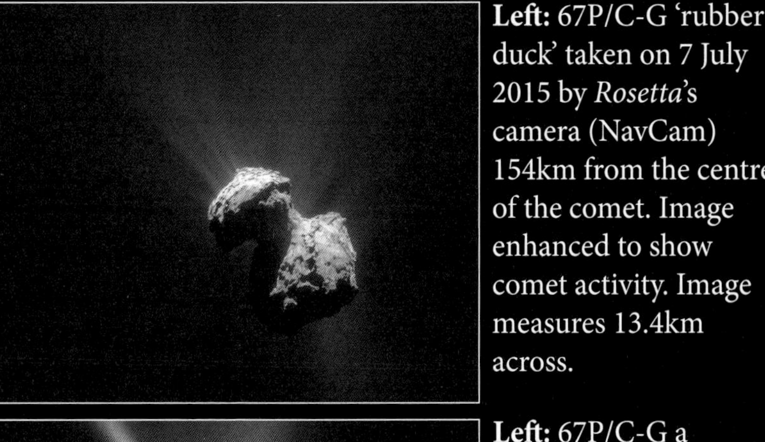

Left: 67P/C-G 'rubber duck' taken on 7 July 2015 by *Rosetta*'s camera (NavCam) 154km from the centre of the comet. Image enhanced to show comet activity. Image measures 13.4km across.

Left: 67P/C-G a few hours prior to perihelion on 12 August 2015, taken by *Rosetta*'s OSIRIS narrow-angle camera from a distance of about 330km. Comet activity is clearly visible.

Below: Four images of Jupiter and the night-side impact of a fragment of Comet Shoemaker-Levy 9. Taken by NASA's *Galileo* spacecraft on 22 July 1994.

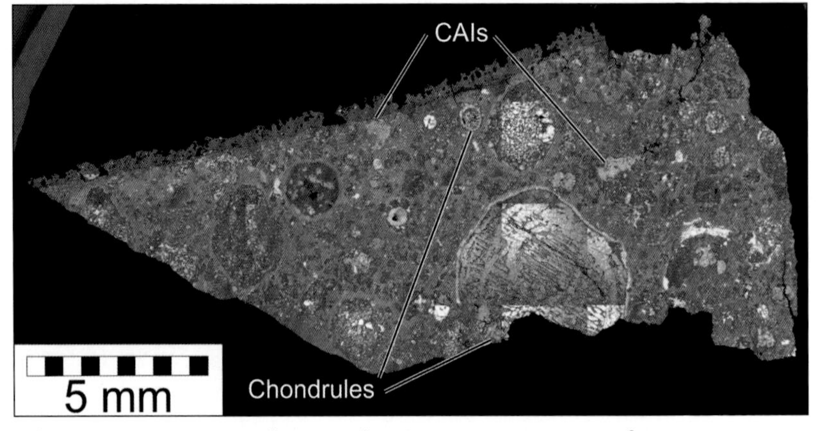

Above: X-ray map taken at the Open University of meteorite RBT 04143 showing chondrules and calcium-aluminium-rich inclusions (CAIs). Colours: red for magnesium, green for calcium and blue for aluminium.

Left: NWA 6727 chondrite meteorite showing the rock interior on one broken surface. A thick fusion crust is visible with a webbing of contraction cracks and flow lines.

Below: NWA 7705, HED achondrite, eucrite polymict meteorite that has been cut and polished to show a dark-grey interior. A shiny fusion crust with contraction cracks is present.

match perfectly with Earth's water. This was an interesting discovery, and one that was used at the time to suggest that Kuiper Belt comets could represent the source of Earth's water, having collided with Earth during the Late Heavy Bombardment (LHB) and parachuted in their volatile inventory. However, the *Rosetta* mission taught us that comet 67P/C-G, which is a Jupiter-family comet also originating in the Kuiper Belt, has a high D/H ratio, three times that of Earth's. This is not to say that the measurements of 103P/Hartley 2's water were wrong. The results show instead that you can't judge one comet by another, even those that grew up in the same neighbourhood. The newer *Rosetta* results showed that even if most of the comets to hit Earth were like 103P/Hartley 2, and only a few were like 67P/C-G, then the composition of Earth's water still couldn't be accounted for. Any 67P/C-G-type comets impacting the Earth, with their high D/H water, would act to raise the Earth's D/H, even if only by a little. Since scientists have figured out how to measure D/H very precisely, this is something that they could detect.

But what about the Oort Cloud comets? Even though they now reside much further out in the Solar System compared with the comets in the Kuiper Belt, they are thought to have originally formed closer to the Sun, near the orbit of Jupiter. As such, Oort Cloud comets might represent a promising source for Earth's water. Of the Oort Cloud comets that have been measured to date, including the famous Halley's Comet, the D/H ratios do not match that of Earth, being around twice as high as terrestrial water D/H. So, Earth can't tolerate too many of these objects hitting it either for it to still retain a low D/H. If Oort Cloud and Kuiper Belt comets didn't bring in all of Earth's water, it seems that another source is required. The next obvious place to look is the asteroids and we know that many of these also collided with Earth in the past.

You'll recall that the asteroid belt is composed of objects that formed at a range of heliocentric distances but that these locations, in general, were closer to the Sun than the comets. Except of course for the asteroids in the outer belt that might, in fact, just be comets that were flung towards the Sun from further out in the Solar System. As such, we wouldn't expect all the asteroids to contain water of the same composition, so we should be a little careful not to group them together. Therefore, to decide whether asteroids delivered Earth's water, scientists can't just look at one or two of them.

The D/H ratios of carbonaceous chondrite meteorites – those that originate from asteroids located on the outer edge of the asteroid belt and are rich in water – have been measured extensively and happen to show very similar values to Earth. This is an exciting result suggesting that they could represent the source of Earth's water, but it is also a surprise and rather intriguing. It seems that they aren't very similar to the comets after all, as they don't share the same high D/H. However, since not that many comets have been measured in relation to the number that exist in the Solar System, any conclusions based on these few measurements can only be tentative.

With all this confusion, all that can be said for sure is that the water on Earth most closely resembles the bulk composition of the carbonaceous chondrites, and their parent asteroids, and at least some of the comets. There is also a strong possibility that some interstellar water survived the birth of the Sun and formation of the planets such that it remains trapped in planets, asteroids and comets, in varying degrees, to the present day. The water on Earth today may be the result of a complex mix of ancient interstellar water (which is older than our Solar System) and newly formed solar water (which was formed in our Solar System, being 4.6 billion years old), all or some of which may have been delivered by comets and asteroids impacting the Earth. It's also possible that Earth's water was delivered in just a few large impacts of comets and/or asteroids. In which case, the D/H ratio of those individual bodies would have been very important in controlling the overall composition of Earth's water. Alternatively, the composition of Earth's water may be a mixed cocktail of contributions from many different objects that collided with Earth and brought with them their unique D/H ratios.

Clearly, it is imperative that the D/H composition of more Solar System objects are measured to enable us to understand the true distribution of water in the Solar System and to make firmer conclusions as to where Earth's water originated. The current findings strongly suggest that Earth's water, or at least the vast majority of it, didn't come from the outer disc comets that are, seemingly, characterised by water with high D/H. This doesn't mean that many comets avoided colliding with the Earth, though. It's almost certain that comets impacted our surface in the past, but the majority of them may have either vaporised away on impact, such that their volatiles were lost to space, or they hit the Earth too early, such that their volatiles were later stripped away by another impact of a different object. The possibilities are endless, but with more measurements in the future scientists will continue to unravel the complicated history of water on Earth.

Organic matter on Earth and in space

As we know, it wasn't just water that was brought in with the bombardment Earth and the inner planets received from comets and asteroids during the LHB, but also carbon, and many compounds made from it. The carbonaceous chondrites, after all, certainly didn't get their name because of a lack of carbon! It's not just the chondrites, though: comets also contain a great deal of carbon that they collected up from the early solar nebula. As we saw in Chapter 4, individual cometary IDPs can be composed of 80 per cent carbon and comet 67P/C-G, visited by the *Rosetta* mission, has a surface as dark as toner ink, indicating a high level of carbon. A comet such as Halley is said to contain organic matter equivalent to around 10 per cent of Earth's current biomass.

Scientists can't be sure whether comets and asteroids, and their inventory of volatiles and organic matter, survived impact into Earth. However, because some of them are so organic-rich, potentially only a small fraction of the organic molecules they transported would need to have survived impact to have had a significant influence on the chemistry of the Earth. When we look at the tiniest of comet particles that reach us, those of the IDPs, the fact that they rain down to Earth continuously, even today, suggests that even without large cometary impacts to the Earth, we might still have gradually and gently received enough organic matter for life to begin. Furthermore, IDPs are packed with volatiles, so they could also have been useful in the delivery of water. So, is there anything about the structure or chemical make-up of the organic material on Earth and in space that can help us to work out where we came from?

Carbon occurs in many different varieties in comets and asteroids, including in inorganic forms such as silicon carbide and even graphite. These are thought to have condensed from very high temperatures in the atmospheres around other stars that existed before our Sun (referred to as 'presolar' grains). Although these are very interesting grains in their own right, we will concentrate here on the more common type of carbon in these objects, which is in the form of organic matter. This can be present as 'free' molecules – represented by hydrocarbons, amino acids and carboxylic acids – or as macromolecular material, composed of more complex, highly cross-linked networks of carbon.

Amino acids in space

It is the amino acids that you'll hear about most frequently, because of their importance as the building blocks of proteins and their obvious role for life. A key feature of amino acids is that their molecules come in two varieties that are mirror images of each other. This feature is known as chirality and the molecules are referred to as 'left'- and 'right'-handed. The molecules contain the same atoms, but they are arranged in a different way such that they can't be superimposed on each other, just like your hands when held the same way up. Life on Earth is known to use exclusively the left-handed form of these amino acids, but scientists don't understand the reason why the left-handed amino acids were, and are, favoured here. This is especially important when scientists also know that abiotic reactions – physical reactions that make amino acids but not involving biology – will produce equal numbers of left- and right-handed forms, something that is known as a racemic mixture. Investigating the handedness, or chirality, of amino acids in space can help to shed light on this problem.

The Murchison meteorite, an over-100kg (220lb), organic-rich meteorite which fell in Australia in 1969, happens to have been a very useful Solar System object in terms of furthering our knowledge about organic matter in space. As discussed previously, not only were many amino acids discovered in Murchison, but they were found to be present in both left- and right-handed forms. However, the remarkably interesting part about these amino acids is that they were found with a slight excess of the left-handed form. This was even the case for the non-protein amino acids: those not used in protein production and which are not encoded with genetic code in any living organism that we are aware of so far. This suggests that left-handedness is an abiotic feature; that it's most likely to be presolar (formed before our Sun) and inherited from interstellar space. Murchison is not the only meteorite to show this excess. The more space materials scientists study, the more they discover that share this similarity. A more recent meteorite fall in 2000, called Tagish Lake, was found to contain about four times as many of the left-handed variety of amino acids as the right-handed. As further evidence, some of the amino acids in Tagish Lake, namely aspartic acid and alanine, were also found to be enriched in the isotope carbon-13. This supported the idea that they formed by a non-biological process, since organisms preferentially use carbon-12 over carbon-13.

But how can an excess of left-handed amino acids be created in space when it is known that abiotic production makes racemic mixtures? Here comes the science again. The cause of an initial bias towards left-handed amino acids is thought to be related to the effects of polarised radiation – radiation produced from things like neutron stars and black holes. Such radiation can act to preferentially destroy one form of amino acid. In the case of our Solar System, this happened to be the right-handed form in organic matter that was sitting on ice and dust grains in the interstellar medium prior to Solar System formation. These grains, with their reduced amount of right-handed amino acids, were then incorporated into comets and asteroids as they formed from the early cloud of solar nebula material. As such, these objects would have inherited the, albeit small, left-handed bias.

A similar mechanism might be at play elsewhere in the Universe, but it could instead have a preference for saving the right-handed form instead of destroying it, depending on the wavelength of the polarised radiation at work. This could mean that, if life is out there somewhere, it might function on the mirror image of amino acids to the life in our Solar System. How this would impact the biological processes and functions of these systems is unclear, though. My younger science-fiction-fan self gets excited about the prospects of this!

A problem, however, is that this mechanism is thought to be capable of introducing only a small bias in the amino acids, of potentially just a few per cent, whereas some of the left-handed amino acid excesses measured in meteorites are much larger. Another method is clearly required to explain how an initially small bias might have been further enhanced on, say, an asteroid – to account for the excesses measured in meteorites – where biology can't have played a role. This is where water flows back into the story. Interestingly, a study comparing the water abundance in chondritic meteorites with the excess of the left-handed amino acids in the same rocks found that those asteroids with higher water content also had a larger excess of the left-handed form. Liquid water is known to have circulated through asteroids, resulting in aqueous alteration as evidenced by chemical and structural changes to the rock. Although it can't have created the initial left-handed excess, it seems likely it could amplify it in some way once it's there, but the exact reason *why* is still being investigated.

Either way, once an excess of an amino acid has been created in space, however small, then it is more likely that life will utilise the more common type of molecule present, which in the case of our Solar System is the left-handed form. On Earth, the abundance of left-handed amino acids has, therefore, been further amplified by the fact that all life on Earth uses them and the preference has been passed down throughout evolution.

Despite the difference in the size of the excess, the fact that an excess of the left-handed form of amino acids can be found in extraterrestrial objects and on Earth doesn't seem like it could just be a coincidence. The similarity between these objects supports the possibility that life on Earth originated from space. Furthermore, if life started here, then it seems that we might be able to expect there to be, or to have been, life on other planets, even if it is still a bit of a leap to turn basic organic matter into even the most primitive life forms.

If somewhere in our Solar System there existed a place that had some water at just the right temperature, then it might be possible. But it will take a few more years and some more sample return space missions to fully investigate amino acid chirality ('handedness') elsewhere in our Solar System. There was an instrument on the *Rosetta* lander that was due to measure the chirality of amino acids on a comet for the first time. However, it unfortunately failed to perform its tests because it needed a rock sample from the *Philae* lander drill which couldn't operate after the unexpectedly bouncy landing of the spacecraft. This just shows that even over 20 years ago, when the *Rosetta* mission was being planned, scientists had their hearts set on measuring amino acids in space. So, we will have to wait patiently for future mission data to see if the amino acids in comets also share the left-handed preference observed on Earth and in meteorites.

How life took hold on Earth, and from which source, remains a mystery, but the important thing is that all the right ingredients for life could have been delivered to Earth from comets and asteroids, or small pieces of them, even if Earth clearly still had a bit of work to do to evolve all the varied life forms that have existed since. As for life elsewhere in our Solar System or Universe, surely it is inevitable.

CHAPTER 6

Taking the Science to Space

Having established that the distances to even the closest comets and asteroids are vast means that visiting them in their natural habitat is technically challenging, financially costly, time consuming and requires patience, as well as technical expertise. Of course, some of these objects come a little closer to Earth as NEOs, which makes them slightly easier and cheaper to reach as they transit the inner Solar System after they've been diverted from their normal orbits in the Kuiper Belt, Oort Cloud or asteroid belt. Studying space objects in the near-Earth environment still has its challenges, but their relative proximity to Earth has some benefits, as they are easier to get to with space missions and they can also be studied using less powerful telescopes, including amateur ones. Nevertheless, intercepting NEOs using spacecraft is still technically demanding because these inner Solar System visitors arrive on trajectories that are challenging to approach with a spacecraft, and at speeds that are hard to achieve with a direct launch from Earth. It certainly isn't a case of launching a spacecraft from Earth and heading in a straight line to meet the asteroid or comet. Often it will take a space mission many years just to catch up with an orbiting NEO, because it has to perform gravitational slingshots around the rocky planets and the Sun to build up enough speed, and to get on to the correct orbit as the object it's trying to catch. In fact, on its 10-year journey to catch up with comet 67P/C-G, the *Rosetta* spacecraft performed three gravitational slingshots of Earth and one of Mars after launch in order to gain enough speed to catch up and approach the comet from behind, before braking to cruise alongside it at the same speed.

An added issue is that comets and asteroids are small in comparison to planets, so they have very little gravitational attraction. This makes approaching and orbiting around one no easy matter, requiring a powered craft capable of manoeuvring itself to achieve an intercept without the benefit of a large gravitational pull from the object it's approaching.

Another of the many complications of approaching NEOs is that comets, in particular, but also some asteroids depending on their composition, tend to be very active in the inner Solar System. If the object is rich in volatiles, it will affect how much material streams off it in the form of rock dust particles and gas as the volatiles inside it are gradually heated up as the object approaches the Sun. The dusty activity produced by the object is enough, by itself, to bombard and cover space instruments in dust, rendering them useless in this situation. Solar panels, for example, obviously won't work very well if they are covered in dust as it blocks the Sun's rays. Some high-speed particles could easily even puncture spacecraft panels or instruments as the process of dust leaving a comet isn't always gentle – some dust leaves its parent body rather explosively in a powerful jet.

Despite all these potential dangers, there have been some key space missions that have taken on the challenge, and risks, of approaching comets and asteroids, and they have all produced spectacular results.

Approaching comets

There have been over 10 comet missions to date. The first, *ICE*, visited comet 21P/Giacobini–Zinner in 1978, with its closest approach at over 7,800km (4,800 miles). This distance sounds quite far, but we have to remember that this mission was very early in the history of space exploration. More recent missions, such as *Giotto*, which launched in 1985, have made closer approaches, so close that the spacecraft was actually hit by some of the high-speed debris streaming off the active comet. The impact of this debris destabilised the spacecraft for over 30 minutes until control was regained. The 2005 *Deep Impact* space mission tried a slightly new approach: it not only visited comet Tempel 1 but also blasted an impactor into the side of it with the intention of excavating a hole to study the interior of the comet. A great deal of dust was thrown out from this impact – considerably more than expected. In fact, there was so much dust that a cloud of dusty debris shrouded the view of the crater that had just been excavated. Even though scientists didn't get a good view of the comet interior at the time, they still learnt from this experience, finding that the comet was dustier and contained less ice than they were expecting. However, it was unfortunate that they were unable to gain more information about the insides of the comet during this mission.

Overall, most missions haven't been able to approach comets at very close range. However, *Stardust* passed sufficiently close to comet Wild2 to collect physical samples for return to Earth in 2006, a ground-breaking success that has not been repeated so far. *Stardust* has been a hugely important space mission for aiding our understanding of the composition of comets and, in turn, our appreciation of how the Solar System formed. So much so, in fact, that I have dedicated the whole of Chapter 7 to the mission.

There is only one mission that has orbited a comet, travelled alongside it to experience its cometary environment for a prolonged period of time and dropped a lander onto its surface – the *Rosetta* mission in 2014. The plans for *Rosetta* sounded a bit like science fiction at first, but thanks to years of careful planning, and immense amounts of technical expertise, success was achieved, bringing science fiction to reality. The little lander, called *Philae*, which was about the size of a domestic washing machine, performed the first controlled landing on the surface of a comet, that of 67P/C-G. The *Rosetta* orbiter lived with the comet for several years, closely following its every move as it approached and orbited the Sun. Such an achievement involved vast teams of scientists working together just to plan the spacecraft's flight each day during its time with the comet. Unhelpfully for the safety of the spacecraft, but useful for the scientific investigations, 67P/C-G was very active during much of this time, so there was always a risk the spacecraft would be damaged, or even destroyed, by rock particles streaming towards it. Just one of the many risks of attempting to study these ancient objects. Just like *Stardust*, the *Rosetta* mission has been hugely important in furthering our knowledge of the composition and structure of comets, and how they behave when they visit the Sun's neighbourhood. We'll learn a lot more about this mission in Chapter 8.

Approaching asteroids

As for asteroid missions, we've done quite well with those, too. Most have been fly-by missions, but the NASA *Dawn* and JAXA *Hayabusa* missions stand out from the crowd for me. *Dawn* ventured into the asteroid belt to study two protoplanets, or large asteroids – Ceres and Vesta, a double whammy of contrasting objects. Vesta, which *Dawn* visited in 2011, is essentially 'dry'. It is a differentiated object showing signs of surface processes much like the inner rocky planets. Ceres, which was visited in 2015, is 'wet', containing water-bearing minerals and resembling bodies from the outer Solar System. *Dawn* was the first spacecraft to visit these kings of the asteroid belt, getting up close and personal with their surfaces, and it will remain in orbit around Ceres in perpetuity, even after the mission itself has officially ended. It's not just beautiful photographs that *Dawn* returned to us, though. As we saw in Chapter 3, the spectrometer instruments aboard the spacecraft confirmed a link between one of the major achondrite meteorite groups on Earth, the Howardites, and the Vesta asteroid, something that scientists could not have established for certain without visiting the object in space.

With *Dawn* having returned tens of thousands of images of the dwarf planet Ceres, the only dwarf located in the inner Solar System, it's no surprise that we've also learnt a lot about this icy world. The most intriguing results have centred around Ceres' so-called bright spots, which were found to be composed of magnesium sulphates – Epsom Salts back on Earth. But *Dawn* also found that Ceres is not particularly dense, having a density of just 2.09g per cm^3, only double that of water. This led scientists to conclude that it must be composed of at least 25 per cent water. Considering we know it contains some rocky material as well, this means that, because of its relatively large size, it must have more fresh water than Earth. Such information is important because it wouldn't take many objects the size of Ceres to have collided with Earth to account for all the water on our planet.

Moving on to the Japanese *Hayabusa* mission that launched in 2003, despite its failure to return the full amount of material it set out to collect from asteroid Itokawa, it still became the first mission to return asteroid samples to Earth, even if they were just dust sized. *Hayabusa* relied on a 'touch-and-go' style collection, a technique that took it very close to the asteroid surface without actually landing on it. This approach makes a mission much less technically challenging, and hence cheaper, since a spacecraft isn't required to land and relaunch from the object. *Hayabusa* deployed its collector as it reached the surface of the asteroid and the plan was for it to scoop samples of rock. Unfortunately, the mechanism to perform this scoop didn't function properly, resulting in much less material being collected than was planned. Thanks to advanced laboratory instruments back on Earth, and the technical skills of scientists, the tiny dust grains that were collected from Itokawa – grains that were sometimes just one hundredth of the thickness of a human hair – have been analysed to discover more about the composition of the asteroid. Since *Hayabusa* remains as the only asteroid sample return mission to date, a little bit of asteroid dust is certainly better than none at all.

However, *Hayabusa* was quickly followed up by *Hayabusa 2*, the next step in this asteroid-sampling journey that will hopefully build on the successes of the first *Hayabusa* mission by returning more asteroid rock material to Earth. *Hayabusa 2* should collect samples in 2018, but this time will deploy an explosive device that will dig into the asteroid.

Why don't we return more rock from space?

The complex nature of collecting rock samples from an object in space, particularly from a tiny asteroid or comet, shouldn't be underestimated, especially when we consider that in 60 years of spaceflight rocks from only three space objects have been returned – the Moon, Asteroid Itokawa and Comet Wild2. Despite the many space missions to visit, observe and analyse other planets in our Solar System, material has still not been returned from them, including Mars, which has been explored with quite a few different orbiters and rovers to date. The reason, as discussed previously, is that it's technically very challenging to perform not only a controlled landing on a planetary surface but also a controlled launch to return to Earth. In fact, returning material from space can easily double not only the cost but also the duration of a space mission. A sample return mission can almost be thought of as two missions in one.

Most of the comet and asteroid space missions have focused on objects that are in our near-Earth neighbourhood, but others have transited as far as the asteroid belt, some 2–3AU away. This is still a considerable distance for a spacecraft, but it is possible to traverse it in human timescales, albeit with the humans remaining firmly in place on Earth. If we want to travel as far as even the closer of the far-flung comet homes, the Kuiper Belt, the edge of which is well over 10 times further than the asteroid belt, then we are in for a long journey and certainly one on which we wouldn't consider taking humans at the moment.

Rather usefully, occasionally the planets align in a way that can give a helping hand to a spacecraft wishing to embark on such a long-haul journey. Of particular importance is the position of Jupiter. Being such an enormous planet, its gravity can seriously affect the orbit of a spacecraft passing in its vicinity, acting to give a huge gravity kick. This is exactly what happened when the *New Horizons* mission to visit Pluto, the king of the Kuiper Belt, zoomed past after its launch from Earth in 2006. Jupiter gave the spacecraft a colossal boost of momentum that increased its speed by 14,000kph (9,000mph), dramatically cutting its total journey time to the outskirts of the Solar System and making *New Horizons* the first spacecraft to document a Kuiper Belt object. When the mission set off, Pluto was still classed as a planet; it was demoted to dwarf planet later in 2006, not that that made it any less important to study.

New Horizons captured countless stunning images of Pluto's icy terrains, surfaces that we can probably expect some its Kuiper Belt neighbours to mirror. Like many of the objects in the Kuiper Belt, Pluto was an unknown entity, as telescopic studies have been of limited use due to its vast distance from us. When the first images of Pluto's faraway world appeared, even the scientists were surprised with what they saw. The ice composing Pluto's surface seemed to act more like rock than a dormant blob of ice. Pluto appeared to have been active in relatively recent geological history. It shows us that there is still a lot to learn about Pluto's smaller Kuiper Belt neighbours, but this will involve getting a closer glimpse of their surfaces than can be achieved with telescopic study alone. Luckily, *New Horizons* continues to cruise into the Kuiper Belt and it will visit other icy objects at close proximity. Well, at around 3,000km (1,860 miles) away, which is very close in space terms. After it's transited the Kuiper Belt it is due to continue its journey for as long as it has power and is expected to join *Voyager 1* and *2*, both of which launched in 1977 to study the outer planets, in the cosmic hinterlands of interstellar space. It is expected that by around 2026 the spacecraft's radioisotope thermoelectric generator (RTG) power source will have decayed to the point that no further scientific observations will be possible. However, in the intervening years, there is the possibility to learn a great deal more about a rarely visited part of the Solar System.

To date, no man-made spacecraft has made it out as far as the Oort Cloud, some 2,000 to 5,000AU away. Even *Voyager 1*, which entered the interstellar space past the Kuiper Belt in 2012, won't reach the Oort Cloud for another 300 years, and will take 30,000 years to cross it. Unfortunately, even our descendants won't be able to learn about the Oort Cloud from the *Voyager* missions, though, as they will no longer be able to communicate, with their RTG power being used up by around 2025. Hopefully the future will hold some serious space travel developments so that we continue to explore further than we have before.

One of the problems we currently have is that there are no comet missions planned for the foreseeable future. Instead, we must make the most of the comet data we already have, gained from only a handful of comets, to inform our understanding of the many trillions of comets out there. Luckily, there are a fair few asteroid missions in the pipeline, with some of these aiming to return samples to Earth. Despite there being a wait of a few years for these to return, the fact that rock sample is hopefully on its way will keep the sample-analysis scientists busy. They must prepare their laboratories to receive these important pieces of space cargo and work out how they will characterise and study them. A key feature of these future asteroid missions is that some will visit the so-called 'primitive' asteroids – those that are like the classic comets – so it's good news for aiding our understanding of the early solar nebula.

The downside of having visited only a very few comets and asteroids in space is that the ones we've called on may not necessarily be representative of the whole. As we've seen, even two large bodies in the asteroid belt can be completely different to each other – Ceres and Vesta – like chalk and cheese. The second problem is that the more we get to know these few objects in detail, the more our simple understanding of the composition and behaviour of comets and asteroids breaks down. But this can be viewed as a good thing, as it also shows how science progresses and forces scientists to rethink their ideas. Scientists are often happy to admit they don't know the answers, as that's exactly what encourages them to carry on investigating, in search of a solution. Their hypotheses are often proven wrong when new experiments are performed, providing new data that allows for new interpretations. Without space missions, scientists may never build up an accurate picture of our Solar System. If they rely purely on the free space-selected materials that arrive on Earth naturally, the meteorites, or on telescope observations, then they may miss the whole picture. Scientists need to get inside and sample these space objects to truly understand how they formed, and how important they are for understanding where we came from.

Sample return versus space-based measurements

Returning space rocks to Earth is paramount for fully understanding the composition of the various objects in our Solar System and in interpreting how they formed. On Earth, we have access to the best and most up-to-date scientific instruments, and it is not easy to transport these instruments into space to make the same analyses, say, on the side of a comet. The majority of scientific instruments require human specialists to run them, adjusting and refining their settings on a daily basis. This is obviously a problem when scientific instruments are in space and are expected to essentially run themselves with pre-programmed code and no human input. These restrictions mean that many advanced measurements we routinely employ in laboratories on Earth are simply not possible in space. Despite present-day robots being very sophisticated, in most scientific applications they can't replicate the intelligence, experience and problem-solving skills of the human mind. As is the case even with something as simple as a domestic washing machine, robotic instrumentation is good at doing its job in a perfectly controlled environment, but it isn't so good when dealing with unexpected glitches. So, returning rocks to Earth for analysis in our cutting-edge laboratories, with our clever human minds helping out, is often the best way.

Another upside of returning rocks to Earth is that we have an inventory of samples available to us for the future. Then, as instrument advances are made, samples can be re-analysed, or new experiments can be performed that we didn't even know were possible when the samples were collected. For example, the samples from the Apollo missions are still being analysed in laboratories around the world almost 50 years after they were collected, and they will keep scientists busy for many years to come.

The impact of space on Earth

The fact that it is often not feasible, either financially or technologically, to return rock samples from space has driven major advances in space instrumentation – advances that have a huge number of positive knock-on effects here on Earth. Any instrument that is sent into space must be able to withstand the extremes of the space environments it encounters, with radical changes in temperature and pressure often being the two most significant variants. These are factors that do not tend to change for scientific instruments on Earth, as their laboratory environment can be controlled, and overall the changes in pressure and temperature on the surface of the planet are much less extreme than in space anyway.

Space instruments must also be built to withstand the high levels of shock and vibration they experience on launch, so they need to be constructed of sturdy materials. However, other key considerations for space instrument design are size and weight – the lighter and smaller the scientific instruments, the better. One of the major costs of a space mission is the amount of rocket fuel required to launch a payload from the surface of the Earth to escape Earth's gravity and start its journey into space. Therefore, the amount of science that can be achieved on a space mission is severely limited by rocket motor design and thrust. The less fuel that's needed the better, and the more likely a mission is to get funded and to be able to spend money on more exciting things. When *Apollo 11* launched to the Moon in 1969, its total weight on the launch pad was 2,800 tonnes, of which fuel was 2,100 tonnes, or roughly 75 per cent. By contrast, the *Apollo Lunar Module* weighed 16 tonnes, less than 1 per cent of the launch weight.

Of course, sturdy construction materials are generally heavy, and light ones are weak, so this is a key area of instrument planning. Steel is strong but dense whereas aluminium is weaker but lighter. Titanium happens to be a reasonable compromise for many instrument parts, being roughly in the middle of steel and aluminium in terms of strength and weight. Spacecraft panels, on the other hand, are often made from carbon fibre, just like Formula One cars. Space instruments also need to work on a limited power budget, with battery design and usage a key factor in the design.

The result is that scientists who want to send their scientific instruments, or versions of them, into space must think of ways to make them small and light. They must also simplify them so that they are reliable and can work with no, or almost no, human intervention. Even when this is possible, scientists must make compromises and these tend to centre around the quality, or precision, that can be achieved on the measurements, or the number of measurements that can be made. However, there are some major benefits to simplifying a complicated scientific instrument to make it small, light and able to function reliably with limited power supplies. These adaptations can even have major benefits on Earth as well as in space.

We'll take the example of the Ptolemy instrument on the *Rosetta Philae* lander as an example of where a space mission has delivered some really interesting spin-off technologies that have practical applications on Earth. Ptolemy is a 5kg (11lb), shoebox-sized, miniature version of a mass spectrometer housed in a laboratory at The Open University in Milton Keynes, UK. The full-size, Earth-based instrument is easily the size of a car garage. Like its larger Earth-based brother, Ptolemy was designed to measure the so-called light elements, such as carbon, nitrogen and oxygen, but on the surface of a comet rather than in a perfectly climate-controlled laboratory. Not only that, but Ptolemy had to work remotely, pre-programmed to perform its scientific routine without any input from its Earth-based scientists. Ptolemy became the first space instrument in history to collect a mass spectrum on a comet – essentially a graphical pattern of the distribution of ions of chemical elements present in a sample. Ptolemy sucked in the neutral exhaust gases from the comet surface, ionised them (by removing electrons to give them an electrical charge), then analysed them to produce a spectrum to be beamed back to Earth. The data – chemical information about the comet gases – were sent to the Earth-based scientists via the link on the *Rosetta* orbiter that was flying around the comet above Ptolemy. But first this signal had to travel via a receiver in Australia, before being sent to Germany then to a server in France where the scientists could download it. All in all, it seems quite amazing that it ever worked.

In the years after *Rosetta*'s launch, those same scientists who were eagerly awaiting data to come beaming back from the comet – a mixed group of chemists, physicists, geologists, engineers and biologists who had designed and built Ptolemy 10 years prior to this momentous occasion – kept themselves busy. At the same time as continuing research for other space missions, they discovered that the expertise they'd developed, and the philosophy of miniaturising and simplifying technology to address a specific scientific problem, had other potential applications. These were a group of people who were adept at taking science concepts 'out of the lab' and putting them to work in the 'real world' – whether that be on a comet, as they did with 67P/C-G, or on Earth-based problems that they later worked on. Instead of building a cumbersome and costly scientific instrument capable of tackling a whole host of scientific problems (*i.e.* a 'jack of all trades'), they concentrated on building instruments that were light, sturdy and transportable, and capable of doing one thing really well (*i.e.* a specialist).

One of the rather surprising spin-offs the team created was an instrument that could make real-time detection of the chemicals released by *Cimex lectularius* – otherwise known as bed bugs. Why is this useful? Well, bed bugs release chemicals when communicating with each other and finding mates. The instrument developed by the Ptolemy team provides a method for hotels to screen their rooms for infestations quickly and efficiently. The chemicals are essentially the cometary 'exhaust gases' in this respect and can be turned into mass spectra just like the cosmic chemicals. With bed bugs being a major potential problem for every hotel in the world – including top five-star resorts – such easily usable, transportable technology could be invaluable for their future trade. After all, in the modern world of harsh online reviews, a bad infestation of bed bugs could ruin a hotel's reputation. So, even though life hasn't yet been found on a comet, Ptolemy has provided a rather unexpected association between bugs and space.

At the other, cleaner-smelling end of the spectrum, the Ptolemy team worked out that similar technology could be used to 'sniff' for new perfume scents, to help perfumeries scientifically blend the exact fragrance they desire. This might not sound like a very romantic way to make a perfume, but perfumery is big business and being able to reliably quantify and reproduce different perfumes over long timescales means being clever about the production process. Finally, in what might seem another complete surprise, the Ptolemy team applied their know-how to one more problem: testing air quality aboard nuclear submarines. The award-winning technology they developed performs a vital safety role, with submarine air quality being of obvious and crucial importance. It's exciting to think that technology that was designed to work on the side of a comet whizzing through space is also busy working at the depths of our oceans.

However, one of the most successful pieces of spin-off technology from the Ptolemy instrument, and a true translational technology (*i.e.* it came directly from the instrument itself rather than being a newly built technology), was that of a little gas control valve. The scientists had to design a valve specifically for Ptolemy that could control the flow of gas between a high-pressure tank (at 50 bar) into the much lower pressure mass spectrometer and it also had to work as an on–off valve. The valve they produced turned out to be exceptionally good at controlling gas flow and holding pressure when needed, with the result being that the technology was patented and licensed in the years following *Rosetta*'s work and is now available for use in satellites. In a large satellite, 25 of these valves could be required and what's even better is that the valve can be used with corrosive substances, as it doesn't contain any plastic or rubber. This means that the valve offers a specific advantage in satellites employing a 'green' propellant known as HTP (high-test peroxide – essentially hydrogen peroxide mixed with water), which is indeed a corrosive substance but is classed as environmentally friendly because it breaks down to steam and oxygen. The use of the 'Ptolemy' valve can, therefore, revolutionise satellite technology, allowing a move away from the use of hydrazine (a rocket fuel that is toxic and costly), making propulsion systems cheaper and more efficient.

From these examples, it's obvious that if it hadn't been for the success of the *Rosetta* mission, and the need to miniaturise and simplify a huge and complicated laboratory instrument, this novel and diverse range of instruments may never have been developed. The requirements for designing space-appropriate instruments forces scientists to think outside of the box, to push the boundaries of the technology they are working with. The often-unexpected spin-offs are simply a big bonus for many other industries. Some people may question the cost of space missions; they can be very pricey, using a great deal of tax payers' hard-earned money. But the effect that these translational technologies can have on worldwide economies and job creation cannot be denied. After all, the money put towards space missions is spent on Earth – on people – anyway. Space is a huge industry on its own, even without considering the many positive after-effects. Anyway, surely no one can really question whether it's worth exploring space when scientists have been able to land a fully functioning laboratory on a comet whizzing past the Sun, however much it may have cost?

So, which wins? Sample return or space-based measurements? It's a hard one to answer because there are some major benefits and unfortunate downsides to both. In reality, the choice is not really in the hands of the scientists since space mission budgets tend to be small compared with the amount that scientists want to do. Attempting to return samples to Earth will make a mission much more complicated and expensive. It is therefore 'easier' to attempt to make the measurements in space instead of returning samples to Earth.

Nevertheless, space-based measurements are usually not as high quality as those made back on Earth, forfeiting some accuracy and precision. A positive effect is the invention of some unexpected new technologies that have far-reaching benefits for humans. Plus, any new information about space objects is useful since we have, relatively speaking, so little. Still, as a sample-return scientist, and a geologist, I would prefer for all space missions to return some rock samples to Earth to keep people like me busy and happy. Furthermore, for all those millions of people, adults and children alike, who visit museums every year, space samples can inspire a whole new generation of scientists. Most people can't fail to be impressed when they get to hold a piece of outer space in their hands. I know, I've seen it happen many times.

CHAPTER 7

The *Stardust* Mission

It was an audacious plan. Launch a spacecraft and fly it through the majestic but potentially treacherous coma – a stream of ancient Solar System debris – trailing behind a comet as it sails through the inner Solar System. Then return to Earth having captured some precious, cometary, dusty litter. That's not all – with it bring back some grains of dust from outside of the Solar System. But wait, surely this isn't all possible? Yet this is something the NASA *Stardust* mission team made look simple.

The *Stardust* mission launched from Cape Canaveral in Florida, USA, on 7 February 1999. It was a 380kg (840lb) spacecraft that set out with the primary mission of collecting dust samples from the coma of comet 81P/Wild2 (hereafter referred to as Wild2, pronounced 'Vilt 2') and returning them to Earth for analysis. The mission went down in history as the first, and so far only, to achieve sample return from a comet. Being a Discovery-class mission, *Stardust* achieved all its science on a relatively tight budget, too, with the total cost coming in at around $200 million, which included spacecraft design, development, construction and mission operations. In space exploration terms, that's cheap! Discovery was a series of low-cost American space missions to explore the Solar System, fulfilling the vision of its founder, Daniel S. Goldin, to be 'faster, better, cheaper'. From launch to sample-capsule return, *Stardust* took seven years and achieved more than just the sampling of a comet in this time. The spacecraft also performed a fly-by of asteroid 5535 Annefrank on its way to Wild2.

Annefrank was a relatively small asteroid in comparison with those that have been visited by spacecraft before, measuring 6 x 5 x 3km (3.7 x 3.1 x 1.9 miles), and even though *Stardust* didn't get particularly close to it, passing at a distance of around 3,000km (1,860 miles), it is still classified as a close fly-by in space terms. *Stardust* captured some images of Annefrank, observing its general shape and key features, but the encounter was important not just because it glimpsed a never-before-seen space object, but also because it gave the spacecraft and ground crews a rehearsal for the Wild2 encounter a year later, testing key equipment for the first time in space.

Collecting comet Wild2

The *Stardust* team chose Wild2 as their target because it was a relatively 'fresh' comet. Wild2 hadn't visited the inner Solar System many times before, so it hadn't been heated up much, meaning that it held on to most of its ancient volatile materials, having avoided much degassing. Wild2 is a Jupiter-family comet that was originally in an orbit between Jupiter and Uranus, between 4.9 and 25AU from the Sun. In 1974, it passed close to Jupiter and was knocked to a new orbit between Mars and Jupiter, much closer to the Sun at 1.58 to 5.2AU, with an orbital period of 6.39 years. Scientists figured out that Wild2's new orbit allowed them to approach it more easily with a spacecraft, with the only problem being that the comet is not very large, as far as space objects go anyway, measuring just 5.4km (3.3 miles) across. Luckily, since the plan was not to land on the comet, Wild2's small size, and resulting lack of gravitational pull, wasn't a problem as it was decided that samples could be collected in a technologically simpler way by flying through Wild2's coma. Scientists knew that this trail of comet debris should contain all that the comet had collected up when it formed in the early Solar System – 4.6-billion-year-old silicate rock dust and fragile organic matter. However, capturing such precious grains needed careful thought. The spacecraft was due to encounter, and catch, particles in the comet dust stream at 6km/s (14,400mph); this process can be likened to trying to catch a high-speed bullet made of fragile dust, ice and organic material – a supersonic dirty snowball – without destroying it.

Even though collecting the comet dust sounds tricky, to say the least, in reality it was quite simple. The spacecraft passed through the tail of the comet, deploying a specially designed tennis racket-shaped collector that would capture and retain the comet particles after they collided with it. The collector, with an area of 1,000cm^2 (155sq in), hinged out of the sample canister as the spacecraft approached the comet, which it did at just 236km (146 miles) from the Wild2 nucleus. This may sound like a large distance from which to collect comet dust, but in terms of comet tails, which are known to extend up to 160 million km (100 million miles) from their comet in some cases, it's extremely close.

The scientists had to think carefully about what substance they used in the collector to trap the comet particles. A collection material that is too hard would cause the comet particles it encountered to simply bounce straight off, like throwing a tennis ball against a brick wall. A material that couldn't dissipate the pressure and heat of an impact would probably destroy the impacting particles, heating them up and vaporising them, whereas a collection material that was too soft might not survive the harsh space conditions, becoming too pliable, and might not be able to hold on to the impacting particles. The scientists found, having tested various substances in the laboratory that involved firing rock particles at different collection materials, that there weren't many well-known materials that could perform this task. This led the scientists to consider using aerogel, first invented in the 1930s but having been developed for various uses over the following decades. Aerogel is said to resemble 'solid smoke', a gel where the liquid component has been replaced with gas. It has a very low density, just 0.2g per cm^3, and low thermal conductivity, with properties almost like air (it is composed of 99.8 per cent air) and a sponge-like structure. The version NASA constructed for *Stardust* allowed for the capture of high-speed particles without their complete destruction, while preventing major chemical and thermal alteration. It was designed to retain an impacting particle – having decelerated it from 6km/s (14,400mph) to 0km/s in the space of just a few millimetres. Great care had to be taken with the preparation and handling of the aerogel because it is very sensitive to changes in humidity. If it were submerged in water it would absorb its own volume, rendering it the texture of slightly soggy tofu, not ideal for preserving fragile 4.6-billion-year-old dust. The aerogel was used to fill 132 small blocks, around 1cm to 3cm (0.4in to 1in) thick. Each block was lined with soft aluminium foil to ease the removal of the aerogel from the collector on sample return. After encountering the comet dust stream, the tennis racket-shaped collector was designed to be safely stowed away in the return capsule, which was jettisoned from the spacecraft to begin its journey back to Earth in 2006.

Scientists waited for the sample-return capsule's touchdown with some apprehension because, even after all that had been achieved on *Stardust*'s long journey, a successful and safe return to planet Earth was not guaranteed. Part of the anxiety came about because a sister mission of *Stardust*, that of NASA *Genesis* that returned to Earth in 2004 having collected solar wind samples, had a 'hard' landing (which you can read as 'crash' landing). Its parachute failed to deploy during atmospheric entry. The result was that the *Genesis* sample capsule impacted the Earth's surface with more force than it had been designed to withstand, such that it became embedded in the desert floor, not an ideal situation for some precious samples of solar wind. With *Stardust* returning soon after this mishap, the scientists were understandably nervous the same could happen again. Happily, the *Stardust* sample-return capsule parachute opened successfully, performing a perfect 'soft' touchdown. The sample capsule had travelled 4.5 billion km (3 billion miles), a journey that took it around the Sun three times and through the tail of a comet. It became the fastest human object ever to enter Earth's atmosphere, travelling at some 12.8km/s (28,600 mph). The preliminary examination of the returned *Stardust* comet dust samples was carried out by an international team of 200 scientists, which began almost as soon as the sample-return capsule had landed in the desert salt flats of the US Air Force's Utah Test and Training Range in Dugway. The capsule was collected and transported directly to the Johnson Space Center (JSC) in Houston, Texas, for examination, at a laboratory specifically designed to receive the precious and unique space samples.

Collecting interstellar dust

Flying through the potentially treacherous tail of a comet to return rock samples to Earth was not the only sample return the *Stardust* mission managed. In 2000, during its journey to Wild2, *Stardust* opened its tennis racket-shaped collector for the first time, even though it happened to be nowhere near a comet. Instead, the scientists hoped that *Stardust* could collect some pieces of interstellar dust using the reverse side of the same collector that later captured the comet dust. Interstellar dust is produced by stars and exists in the enormous interstellar spaces between them. It was found, from dust detectors aboard the *Galileo* and *Ulysses* spacecraft (launched in 1989 and 1990, respectively), that there is a stream of interstellar dust flowing through the Solar System at very high speeds, around 30km/s (67,000mph). The *Stardust* team decided that they would try to capture some of this elusive material originating from outside of the Solar System. The *Stardust* interstellar collector was exposed to the dust stream for four months before being stowed away ahead of a second collection in 2002. The scientists calculated that this exposure should be long enough to collect some small interstellar grains, which is exactly what they did.

Stardust interstellar grains are important because they are pristine, representing the contemporary dust of interstellar space, where 'contemporary' in astronomical terms means something around 50 to 100 million years old. By studying this dust, scientists can hope to understand details about the processes taking place in our galaxy. Interstellar grains are commonly found in meteorites and IDPs, and so interstellar space can also be investigated using these free samples. However, in meteorites, the interstellar grains are much older, having been hanging around when the solar nebula – and the comets and asteroids – formed 4.6 billion years ago. As such, these older interstellar grains are thought to have been modified and altered during the formation of the Solar System, before being incorporated into the meteorite or IDP parent bodies. This makes them more complex to study and doesn't allow the scientists to see what is going on in the galaxy at the present day, so to speak. The *Stardust* mission offered a unique collection opportunity for free, since the spacecraft was up there anyway, making its way to Wild2.

Analysing comet dust

Once the sample-return capsule was safely installed at JSC, scientists had to carefully assess how successful the sample collections had been. The initial focus of the preliminary examination was on the comet samples, as these were always the main focus of the mission. The question was: how well did the comet dust survive its encounter with the aerogel? It was soon seen that the dust had produced distinctive carrot-shaped burrows into the aerogel collector that were a few millimetres in length, making some of them easy to spot in the laboratory. On closer inspection, however, there was a slight problem. As the particles had burrowed their way into the aerogel it was found that they'd tended to break apart. Little bits and pieces of the dust ended up being deposited all along the inside surface of the burrows. Often at the very end, having excavated its way right into the aerogel, would be a tough piece of rock that became known as the 'terminal particle'. This particle was the hardest, most resistant part of the original piece of comet dust that struck the collector, most often composed of hardy, rocky minerals. It became the easiest part to extract and analyse in the laboratory, simply because it was relatively large and remained intact.

It was the more fragile materials, such as the softer rocky minerals and organics deposited all along the sides of the burrow, that were more challenging to recover for analysis. These early observations about the way the aerogel had trapped the dust indicated to the scientists that it hadn't made the perfect collection medium. Ideally the particles would have been captured and preserved in the exact form that they left the comet nucleus – an agglomerate of different comet components. However, this was probably an impossible task with comet particles that were composed of such a mixed bag of materials – hard and very soft. In fact, what was particularly intriguing to the scientists was how such large, hard, rocky particles were so closely associated with fine, fragile materials in the first place. They weren't expecting discrete large and rocky particles of this sort to be present at all. If Wild2 represented a 'classic' comet, then the scientists were anticipating collecting the typical fine-grained, fragile dust that made up the early solar nebula, just like that seen in fluffy IDPs. The terminal particles, for a start, were large enough to see with the naked eye – much larger than 'comet' dust and more akin to the materials found in asteroidal meteorites. Already the scientists were learning that Wild2 had a great deal to reveal before the proper analyses had even begun.

Rather unexpectedly, the aluminium foil that was used to line the aerogel blocks in the collector, which was there almost as an afterthought to aid in the removal of the aerogel blocks after sample return, had an unintended use as a secondary particle collector. Aluminium foil is certainly not an ideal collection material for high-speed particles, being far too hard and unforgiving for a rock to survive impact intact. This was exactly the case with the *Stardust* foil. The particles that struck the foil didn't survive their encounter very well at all, but they didn't bounce off. Instead, in most cases, the impacting particle was obliterated on contact with the foil – almost completely vaporised – leaving behind a small impact crater. When looking at these impact craters in the foil, the scientists found that they could be grouped based on their shape, with many different appearances seen. Even more interestingly, the foil craters preserved a thin residue of material – a sample of the original impacting particle. The residues were the remnants of the comet particle that had melted on contact with the foil due to the sudden and rapid change in pressure through compression on impact. The scientists quickly realised that the aluminium foil had provided them with more comet samples than they'd planned for. With specialised laboratory techniques, the *Stardust* team could analyse these residues, using highly specialised microscopes to find out which elements they contained. They saw that the shape of the craters was related to the composition of the impacting particle, and from this they could start to work out the overall composition of Wild2 from the mixture of craters they saw. This was an enormous achievement considering that the rock residues were of the order of just 20–100 nanometres in thickness, just one thousandth of a human hair. Such careful analytical work allowed the scientists a greater insight into what the overall composition of the comet was, information that was unexpected yet very useful.

At the same time, the scientists started to study the aerogel burrows in detail and found that they, too, varied in shape. They could indeed be 'carrot-shaped' but others were more 'bulbous', and after careful analysis of the materials left behind in each of these distinctively shaped burrows, the scientists started to see a pattern. Just as with the foil craters, there was a link between the composition of the impacting particle and the shape of the burrow it produced in the aerogel. This allowed the scientists to focus their efforts on different-shaped burrows depending on what they were looking to investigate. For example, particles composed of a higher proportion of 'soft' material, or more volatile materials, would produce a burrow that was squat, or bulbous, because the soft material couldn't penetrate as far into the aerogel. In some cases, soft comet materials might even have exploded as they contacted the aerogel and their volatile components reacted to the pressure and temperature changes. The deeper, narrower burrows, on the other hand, were produced by the harder, rockier particles that could confidently zoom straight into the aerogel.

One of the more remarkable terminal particles that was recovered from its aerogel cocoon very early in the investigation was a grain named Inti, after the Inca Sun God. Inti was an interesting collection of rock minerals that together formed a crystalline mass, or agglomerate, of rock. The chemical and mineralogical composition of these minerals was indistinguishable from some key components found in meteorites – that of the CAIs. Scientists definitely hadn't anticipated finding such a mixture of high-temperature minerals in an apparently 'primitive' sample of a comet originating from the cold edge of the early solar nebula. Soon after the initial characterisation of the particle was complete, it was decided that Inti didn't just *look* like a piece of a CAI, it *was* a piece of a CAI. The composition of Inti indicated to the scientists that it can only have formed very early on in Solar System history, within the first few million years, and probably very close to the Sun, at temperatures well over 1,000°C. Inti looked more like a sample of an asteroid! The presence of Inti in Wild2 was, therefore, somewhat unexpected. However, as they continued their investigations, the scientists discovered that Inti wasn't alone; it was just one of many high-temperature, supposedly 'asteroidal' grains found in the *Stardust* collection.

Nevertheless, of the thousands of cometary grains collected by *Stardust*, of course it was not just high-temperature components that were found. Wild2 was not an asteroid – it behaved like a comet and followed the orbit expected for a Jupiter-family comet. Sure enough, Wild2 also contained the fine and fragile materials normally expected in a comet. In fact, the bulk of the material collected by *Stardust* was composed of smaller particles – nanometre- to micrometre-sized grains – of the type of dust that was expected from the study of samples such as IDPs (see Chapter 4). The stereotypical fragile, fine-grained 'cometary' dust, made of minerals such as olivine and pyroxene, organic matter and some metals, was there, it was just much harder to see. The particles were also harder to remove for analysis, being smeared into and mixed with the aerogel. The scientists worked painstakingly on removing these samples for analysis – they weren't about to give up on the chance of measuring rare samples of a comet just because they were proving tricky to access.

Wild2's organic matter

Using specially developed techniques, the scientists could start laboratory analyses of the *Stardust* organic matter, even when it was still contained within the aerogel. These studies showed that the organic matter resembled that found in IDPs more than that found in meteorites, although it was noted to show some similarities to both. This is a useful piece of evidence supporting the idea that IDPs are indeed of cometary origin, something which can never be known for certain (see Chapter 4). This also means that scientists can confidently continue to use IDPs to explore the composition and formation of comets, in lieu of any further sample-return cometary missions, of which there are none currently on the agenda. IDPs and *Stardust* samples together represent an important inventory of comet samples. While IDPs sample a wide range of comets, it is unknown which exact comets they sample. Whereas, although the *Stardust* samples only come from one comet, it is obviously known that they are from Wild2, allowing them to be placed in context with all the other information scientists have about this object. These differences make the collections very valuable when viewed separately and together, allowing greater insight into the comet-forming region.

An exciting find within the organic matter collected by *Stardust* was the detection of glycine, which is the simplest of the amino acids. This was the first time that an amino acid had been detected in a comet but its presence shouldn't be too much of a surprise since, as we've seen in previous chapters, a wide range of amino acids have been found in meteorite samples. However, it wasn't exactly easy for the scientists to detect the glycine in the *Stardust* samples. In fact, the glycine was particularly hard to identify in the aerogel because the organic matter, being even more fragile and more easily altered than the fine cometary rock dust, was found to have very much melted into the aerogel on contact. Sometimes the organic matter was so heated and disrupted during deceleration that it was encapsulated within pockets of aerogel, making it almost impossible to remove for analysis. Amazingly, however, some glycine was preserved in the aluminium foil and in the aerogel. In particular, after careful, painstaking laboratory investigations, glycine was detected as part of the thin layer coating the base and sides of some of the foil impact craters. Of course, the scientists had to prove that the glycine wasn't present by accidental contamination, introduced by the scientists themselves, or even during the manufacture of the collector before launch. After all, glycine is a very common amino acid found in life on Earth. Luckily, the handy isotopes of carbon came into play and, on analysis, the glycine in the *Stardust* samples was found to show an excess of carbon-13, the heavier isotope of carbon. This showed that it definitely formed in space and not through biological processes on Earth.

Finding interstellar dust

Although it might seem like an easy step for the scientists to apply the same analytical techniques to the reverse side of the collector containing the interstellar dust, unfortunately things just weren't that simple. Interstellar dust is very small, orders of magnitude smaller than even the cometary dust, from just a few molecules to a third of a micrometre across. Scientists estimated that, despite the interstellar collector being exposed to space for 195 days in total, there might be just 45 particles on the whole of the $0.1m^2$ ($1.08ft^2$) surface. The small size of these elusive grains was going to make them hard to find – a microscopic version of a needle in a haystack – and, once found, they were going to be more of a challenge to work with than the cometary particles. Since this search couldn't be carried out with the naked eye, the scientists had to image the entire collector with a camera attached to a microscope. By magnifying each aerogel block, they created a database with many thousands of images that included details such as the tiny burrows excavated by the interstellar dust, just like the comet burrows but smaller. Unfortunately, at this small scale every imperfection on the collector, such as scratches from manufacture or handling, was also magnified, so identifying a small interstellar dust burrow was made even harder.

The scientists soon realised that searching these images for interstellar dust was a monumental task for them to undertake on their own, involving too many hours of work for the relatively small team. Rather ingeniously, they decided to involve the public to help them out, using the 'Stardust@Home' initiative. This was a citizen science project that allowed would-be scientists at home to volunteer in the search for interstellar particles – they aptly named themselves 'dusters'. After an online tutorial and test on how to spot an interstellar dust burrow in the aerogel, the 'dusters' gained access to a virtual microscope connected to the huge database of digital images of the collector surface and they could begin their search. If they identified a potential particle on one of the images, and it was later confirmed by the scientists to be true interstellar dust, the duster got to name the particle. On 14 August 2014, and after many hours of careful searching, the *Stardust* scientists confirmed the identification of the first interstellar dust particles from the *Stardust* capsule. The 'dusters' managed to search over a million images and found particles that were only three picograms, trillionths of a gram, in weight. Not only had the scientists managed to complete their investigations a lot more quickly, but they'd also managed to excite the public about interstellar space – a subject area that many people had never even thought about before.

Of course, finding the *Stardust* interstellar grains was just the first step. The next – the laboratory analysis – was even more complicated. Even though the scientists knew that the analysis of minute dust was going to be tricky, one of the good things about sample-return missions is that there is usually plenty of time – usually years – while waiting for the samples to devote to developing more advanced instruments capable of the required novel analyses. This often means that analyses that wouldn't have been possible at launch become routine by the time of sample return. The other good thing about sample return, of course, is that the samples are stored on Earth awaiting further future improvements in laboratory technology.

So, what did the scientists discover in the interstellar collection? The preliminary laboratory investigations of the *Stardust* interstellar grains indicated that the dust was more diverse in size, composition and structure than had been predicted. It was thought that all interstellar dust should be the same, consisting of single dense particles with the same structure. It was also predicted that the composition of small and large pieces of dust would not differ from each other. This simple picture was proven wrong with the samples *Stardust* provided. Some of the dust was even composed of large, fluffy, agglomerate grains that were totally unexpected. The wide range in interstellar particles captured by *Stardust* indicated that each particle must have its own individual history, rather than it all having formed in the same place at the same time, or in the same way. There is still a lot to learn from the *Stardust* interstellar dust collection; there are many more particles to be investigated and further analyses will be ongoing for many years. The important point is that by sampling the dust transiting the Solar System from interstellar space – a location that is virtually impossible to sample because of its immense distance from Earth – the *Stardust* samples have started to unlock the secrets of the stars. They have allowed the scientists to see what the space between the stars looks like and even to understand how elements are created. For a 'budget' space mission whose focus was on studying a comet, *Stardust* had achieved a great deal more.

Comet or asteroid?

The picture that emerged from the *Stardust* burrows and aluminium foil impact craters was that the particles impacting the collector, which were shed from the comet, arrived as complex agglomerates. They were made from occasional large, hard, rocky grains (such as CAIs or chondrules), encapsulated within, or attached to, an abundance of fine-grained, tiny, fragile dust particles, often mixed with organic matter. It's possible that the organic matter had acted like a glue, loosely holding this complex Solar System assortment together in the comet until the impact forced the mixture apart. As we saw in Chapter 4, organic matter has been reported to act as a glue in cometary IDPs, so such an association was not unexpected for Wild2. However, the association of fine, fragile dust and organic matter with large and rocky particles in the *Stardust* samples was intriguing. It indicated that some of the most ancient materials in our Solar System, which formed at the very highest, and very lowest, temperatures available at the time, had found their way to each other. Such materials were expected to have formed in completely different parts of our Solar System. We're not talking small distances here, either: these high-temperature grains, if formed near to the Sun, must have transited tens to hundreds of AU to meet their eventual cometary host. During this period of Solar System formation, matter was expected to be falling *inwards* and swirling around the young Sun to maintain the angular momentum of the system, so scientists had to reconcile how some grains were clearly going in the opposite direction. This unexpected finding had important implications for models of Solar System formation and evolution and needed to be investigated further.

The results from *Stardust* marked a major turning point in our understanding of the composition of comets, and of their formation. Before *Stardust*, comets were thought of rather simply – they were dusty snowballs – and they were certainly not expected to contain hard, rocky materials formed at high temperatures in the inner Solar System. Such materials were, instead, thought to be restricted to the asteroids and planets, objects that were built in the inner Solar System. As such, comets and asteroids were always considered as discrete entities in space, formed in separate corners of our Solar System and representing very different environments of formation. Wild2 is known to be a comet and, therefore, the samples collected from its coma are definitely of cometary origin. From the collection of some tiny cometary rock samples, *Stardust* showed that the traditional view of comets that had been the accepted consensus for many decades needed a careful rethink.

While scientists still think that the vast majority of comets, including Wild2, formed very far from the Sun, as we've seen, that is not to say that they didn't form over a vast range of heliocentric distances. It's unreasonable to expect them all to have formed in the exact same place. Clearly, some formed a little closer to the Sun and some further away. It is generally accepted that the early protoplanetary disc had a strong temperature gradient – searing, intense heat close to the Sun, yet within 2–4AU the ices of volatiles such as water and methane could form, when the snow line was reached. Nevertheless, the important thing is that the comets must have formed past the snow line, because they are known to contain abundant volatiles in ice form. In this respect, scientists just had to explain how high-temperature 'asteroidal' dust was transported from the inner Solar System in relatively short timescales – probably within a few million years – over huge distances to be incorporated into the comets as they were forming.

Moving the inner Solar System outwards

It was hypothesised that strong jets, or turbulent winds, that are thought to have been active very early on during the collapse phase of the solar nebula, could have been capable of transporting newly formed solids from next to the Sun all the way out to the comets. Such jets were thought to have been active at the time, shooting out along the rotational axis of the system due to an interaction between the strong magnetic field at the time, and the dust and gas falling in towards the star. If they were a feature of the early protoplanetary disc, then they must have been exceptionally powerful, such that they could traverse massive distances against the general flow of material falling inwards. Just like a desert wind picking up sand, it is proposed that these windy jets could transport chondrules and CAIs, or fragments of them, rapidly flinging them out of the inner Solar System to be deposited in the cold comet-forming region as the winds lost momentum. Once they were deposited, the particles must have been rapidly incorporated into the newly forming comets, before they had a chance to fall back in towards the Sun.

Such models are attractive as they can account for the presence of particles like Inti within Wild2, but the timing for all this to happen was very tight (on geological timescales anyway), occurring within the first few million years of Solar System formation. While these winds seem to represent the prevailing model to explain the *Stardust* results, there are a few problems. When scientists looked in detail at these windy jet-transport models, they found that they were only capable of transporting very small solids – dust-sized material up to about 2 micrometres across. Anything larger and heavier probably couldn't get picked up and transported. Indeed, many of the high-temperature solids found in the *Stardust* collection are small enough, but others are larger, even if they are still only tens of micrometres across. While these models are still being refined, they might yet be able to explain the observations, but equally, perhaps there are some parts of the early solar nebula that scientists still don't understand well enough. We certainly don't have all the answers yet but, fear not, there might be a solution lurking out there in a computer model soon, or in another comet particle.

However, it's always good to keep an open mind as there are, of course, alternative possibilities to account for the puzzling results presented by *Stardust*. Another scenario that has been presented questions the assumption that high-temperature solids must be formed within the inner Solar System. Instead, it is suggested that early on in Solar System history, around the time of planetesimal formation, it was possible for localised hot regions – with temperatures in excess of 1,000°C – to exist in the outer Solar System. This may sound unlikely as we know that the ambient temperature in this region is near enough absolute zero (-273°C). However, the outer protoplanetary disc is thought to have been gravitationally unstable at the time and, as such, capable of forming so-called 'giant planetary embryos', such as Jupiter. It is thought possible that some of these 'giant planetary embryos' were capable of processing solar nebula dust at high temperatures to make similar-looking solids to those formed at high temperatures next to the Sun, *i.e.* chondrules and CAIs. If these giant planetary embryos were broken apart when migrating in towards the Sun then they could have released and deposited their gas and high-temperature solids into the surrounding cold disc, where they would have been available to be incorporated into the comets. While this idea may sound a little wild, it doesn't mean that it isn't a viable option. There is still a great deal that is not understood about the first few million years of Solar System history. The question remains: can scientists tell apart solids formed in hot, giant planetary embryos at large heliocentric distances and those that formed within the inner Solar System? Clearly, this is an area that needs further investigation and the *Stardust* samples provide an important inventory with which to answer such questions.

As you'll see, it's currently not possible to find a single theory on which all scientists will agree, and which can account for the new results that have come from this important comet mission. However, analysing the *Stardust* samples is a painstaking process, and there are lots of individual particles to focus on, even if they are very small. These tiny rock grains may hold more clues as to how, and where, the high-temperature material within Wild2 formed, and how it came to be so closely associated with low-temperature materials within the comet. There is always the possibility that Wild2 is a unique and strange comet, unlike any other, but even if this was the case then scientists still need to account for the association of high- and low-temperature phases that are so intimately linked within this single space object. There are so many comets out there though, that it seems likely Wild2 was not alone, and that many other comets must share a similar, even if not identical, Solar System history.

The impact of *Stardust*

Despite all the confusion over which Solar System ingredients should be expected in comets, and whether a 'classic' comet is really a thing, what can be said for certain is that the vast majority of the ingredients in comets haven't ever experienced high temperatures, during their formation or since. Comets retain and preserve their very important inventory of ancient, and pristine, solar nebula materials. The analysis of comet samples is, therefore, beneficial for our understanding of the early solar nebula, since they are the best way to access these rare, ancient materials. The fact that comets are also found to contain solids that are more commonly associated with asteroids has led to some confusion over their formation, but it doesn't make them any less useful for understanding the early Solar System. Quite the opposite, in fact, as they record other processes that were taking place, which scientists didn't know about before.

The new findings have led to a debate over whether there might exist a continuum of compositions between asteroid and comet: from fully fledged icy, dusty ancient comet – the classic 'dirty snowball' – to rocky high-temperature asteroid, with everything in-between being a mix of the two. This might mean that in the very middle there is an object that is as much asteroid as it is comet. Such an idea is attractive to account for the many varied objects that have been observed so far, and the fact that there is certainly some overlap in the ingredients that can be contained in comets and asteroids. After all, we have seen that some of the asteroids can contain very primitive ingredients, such as fragile organic matter and fine-grained dust, and display 'cometary' activity like the production of a tail. However, it might be unnecessary to assume there is a continuum of compositions. Instead, it seems more likely that each group of space objects, whether you take the asteroids or the comets in isolation, can contain some of the same ingredients as each other, but that they fundamentally formed in different places and environments.

It is by studying comets and asteroids using IDPs, meteorites and space missions (whether they return samples or not) that we can begin to answer some key questions about the formation of objects within our Solar System. Of course, there is much still to understand, not just about comets and asteroids, but about the formation of the entire Solar System, including the other planets. However, without understanding the starting ingredients where everything began – by analysing the most pristine and primitive space samples available – scientists can't hope to unravel the more complicated asteroidal and planetary histories. The *Stardust* mission, while being a huge scientific and technological success, has clearly presented scientists with more questions to investigate, but this is a good thing as they are now closer to knowing the answers. Another important feature is that the mission opened debates in the scientific community and challenged long-held beliefs concerning important details about the Solar System's formation. At the same time, it delivered precious samples that are available on Earth in perpetuity; they might even be able to answer questions scientists haven't thought about yet. For such a cheap and quick space mission, the first to attempt the sampling of a comet in space, *Stardust* surpassed many expectations, having achieved so much with so little. It turned our understanding of comets, and of their formation in the early Solar System, upside down. Future sample-return missions to comets and asteroids will build on these findings and scientists will be able to better direct their scientific questions in the light of what *Stardust* has shown. It may seem that very slow steps are made in our quest to find out more about these space objects, but when these infrequent steps are taken, they tend to result in big findings that can't be ignored. We'll see more about this in Chapter 8.

The future of *Stardust*

Even after the return of the *Stardust* sample capsule to Earth, the mission itself was far from complete. Despite all it had achieved, the rest of the spacecraft entered hibernation to conserve energy and continued its space journey as *Stardust-NExT*. During this mission, it intercepted and explored comet Tempel 1, another Jupiter-family comet that had been the literal target of the *Deep Impact* mission in 2005. This marked the first time that a comet nucleus had ever been revisited and gave scientists a chance to image the crater created by the *Deep Impact* ballistic impactor that had excavated a hole in Tempel 1, as we learnt in Chapter 6. The *Stardust-NExT* encounter with Tempel 1 allowed scientists to study the geology of the comet, seeing how material appeared to 'flow' like a liquid on the surface. When it completed this work in 2011, Tempel 1 became the most mapped comet nucleus to date. Phew. I think you'll probably agree that *Stardust* really did achieve a great deal with its $200 million budget.

CHAPTER 8

The *Rosetta* Mission

The ESA *Rosetta* spacecraft had a lot resting on its wide, solar-panel shoulders when it launched at the start of its mission to catch up with a comet in space. Firstly, it had a long and lonely journey into deep space to contend with, requiring it to enter hibernation for a number of years to save energy. There was no certainty that it would wake again. After an immense, 6-billion-km (4-billion-mile) journey, *Rosetta* was set to make space history by entering orbit around a comet, and then landing a fully functioning laboratory lander, *Philae*, on to it. Two things that had never been attempted before. The comet in question, 67P/C-G, hadn't been studied in detail before *Rosetta* approached, so very little was known about its small, primordial features. Even after the decade-long journey for *Rosetta* to intercept 67P/C-G, as well as the decade of mission planning prior to launch, *Rosetta* scientists and engineers had to wait until the spacecraft was very close to the comet before they could start planning the mission to orbit and land on it. They simply had no clear idea what 67P/C-G even looked like before they got to it. Because it was unknown what type of cometary surface they would encounter, *Philae* was designed to be capable of successfully landing on anything, from as hard as concrete to as soft as candy floss. These challenges, and ultimate successes, make the *Rosetta* mission, in my opinion, the most impressive and awe-inspiring of the modern space era.

Chasing a comet

It's been over 20 years since the *Rosetta* mission was approved as part of the ESA's Horizons 2000 Science Programme in November 1993. The three-tonne spacecraft was made up of the *Rosetta* orbiter, with its 14m (46ft) long solar-panel wings and 11 instruments, and the *Philae* lander, which piggybacked on *Rosetta* with its own suite of nine instruments. This plucky spacecraft launched successfully from Kourou in French Guiana, South America, in March 2004 aboard an Ariane 5 rocket. However, even the very powerful Ariane 5 couldn't provide enough initial velocity at launch to enable *Rosetta* to transit directly to its target comet without the need to gather more momentum in space. *Rosetta* was required to match the exact orbit and speed of the comet it was chasing, after it had sneaked up on it from behind. The result is that *Rosetta*'s journey resembled something like a cosmic pinball game as it performed multiple close fly-bys of planets – the Earth in 2005, 2007 and 2009, and Mars in 2007 – four orbits around the Sun and two transits through the asteroid belt. These manoeuvres allowed *Rosetta* to gain gravitational energy, speeding it up and pushing it onto a new orbit each time, edging it ever closer to the dusty snowball it was chasing.

The mission was named after the Rosetta Stone, an inscribed, volcanic slab of rock that had revolutionised our understanding of ancient Egypt. The rock can be seen today at the British Museum, London, UK. The Rosetta Stone has carved inscriptions in three different languages that provide a way to decipher the mysterious ancient written language of hieroglyphs, which in turn unlocked secrets about a lost culture. The *Rosetta* space mission was given this name in the hope that it too would unlock secrets – the ancient secrets of our Solar System – by catching up with a comet in space. The *Philae* lander was named after the Philae obelisks that were used along with the Rosetta Stone to decipher the Egyptian hieroglyphs, one of which can be seen today at Kingston Lacy house in Dorset, UK. The *Philae* lander was used alongside the *Rosetta* orbiter, being deployed to the comet surface to unlock the secrets held inside.

For the *Rosetta* mission, however, orbiting and landing on a comet weren't enough, and during its long journey it had the opportunity to perform two close fly-bys of asteroids. Much as we'd seen with the *Stardust* mission, these fly-bys gave *Rosetta* a chance to spread its wings and test out a few instruments, and its communication systems with Earth. This was a great opportunity for the anxious scientists and engineers, because there was no guarantee that everything would work as planned once the spacecraft had begun its journey, especially after a powerful launch and long journey through cold and radiation-filled deep space.

The first of the two main-belt objects glimpsed by *Rosetta* was asteroid 2867 Steins in 2008, which it passed at a distance of just 800km (500 miles). The seven-minute-long encounter, in which Steins was beautifully illuminated by the Sun on the side facing *Rosetta*, produced some very clear and bright images as it was tracked by *Rosetta*'s cameras. The distinctive shape of the 4.6km (2.85 miles) wide object, and the way it appeared to glint, led to it being named the 'diamond in the sky'. Consequently, during later studies of the object, the craters observed on its surface were named after gemstones. When *Rosetta* approached its next main-belt asteroid, 21 Lutetia, in 2010, it was much further away from it, at a distance of around 3,000km (1,860 miles). However, *Rosetta* was still capable of capturing high-resolution images of Lutetia, as the large approach distance was mitigated by the fact that Lutetia was itself a large object, being 100km (62 miles) in diameter.

Rosetta also attained some chemical mass spectra from Lutetia. When the results came in, at first the mission scientists thought that the *Rosetta* instruments had failed to analyse the asteroid at all, being too distant to 'sniff' its gases. Instead, they thought they had simply sniffed the spacecraft itself, or rather its exhaust gases. However, it turns out that even the great distance between the asteroid and spacecraft was not too large for *Rosetta* to work successfully, as after careful examination of the data the scientists saw that *Rosetta* had picked up the scent of the asteroid after all. This was fantastic news as it meant that the scientists could breathe a sigh of relief, knowing that their instruments had worked correctly and they'd obtained some fresh data.

The Lutetia encounter, however short, still allowed the scientists to learn a great deal about the asteroid. Its heavily cratered surface revealed a complicated and tumultuous history, with crater ages spanning from 3.7 billion years to within the last few hundred million years, relatively recent on geological timescales. One of Lutetia's craters was found to be 57km (35 miles) in diameter, which numerical simulations indicated was formed by a 'projectile' impact (probably another asteroid), itself at least 7.5km (4.6 miles) in diameter. The cratering history, number and ages of craters in different regions across Lutetia's surface helped the scientists to see that the asteroid appeared to be composed of different geological terrains that must have formed over a wide span of time. Lutetia was not an object that had experienced a simple formation history. It was also found to be a differentiated asteroid, containing a core and primitive crust, something which was unexpected for an object of this *relatively* small size.

After this 'leg-stretching' phase of the *Rosetta* cruise, the spacecraft was set to travel much further away from Earth, out into the cold and dark of deep space. The scientists knew they would need to shut down most of the spacecraft for a significant portion of its upcoming journey, because its solar panels would receive too little light to keep it fully powered so far from the Sun. *Rosetta* entered deep-space hibernation from June 2011 to January 2014, just keeping enough systems running to stop it from completely freezing to death. The next big hurdle for *Rosetta* was reactivation from hibernation – for it to be awoken from its deep-space slumber – something that couldn't be guaranteed.

Luckily, *Rosetta*'s internal alarm clock, which had been ticking since launch all those years earlier, went off at 10 a.m. GMT on 20 January 2014. This triggered a complex series of events to bring the spacecraft out of hibernation, which it managed successfully, waking from its coma without the loss of any systems. This was lucky because the *Rosetta* teams had no way to ruffle the spacecraft's covers from Earth. With the alarm clock having gone off, *Rosetta* activated its core systems, such as heaters and avionics, to slowly bring the whole of the spacecraft back to life. It fired its thrusters to slow down its rotation and pointed its solar arrays towards the Sun to begin charging up its batteries again. *Rosetta*'s on-board navigation system – the star tracker – became active to work out the location of the spacecraft. The spacecraft then switched on its transmitter and all the systems were checked over the following few days. However, the science teams would have an anxious wait to find out how their instruments were performing as the checks on these took place over the following few months. Once all checks were completed it was found that *Rosetta* appeared to be in good working order and was ready to begin the most crucial stage of the mission.

Why comet 67P/C-G?

Comet 67P/C-G was the lucky object chosen by the *Rosetta* scientists to be investigated in detail, to force it to reveal its 4.6-billion-year-old Solar System secrets and become one of the most well-known comets in existence. The comet was first observed by, and named after, Soviet astronomers Klim Ivanovych Churyumov and Svetlana Ivanovna Gerasimenko in 1969. 67P/C-G originated in the Kuiper Belt, where it had a perihelion distance of around 2.7AU, until February 1959 when a close encounter with Jupiter diverted it inward, where it remains today, giving it a perihelion of just 1.3AU. 67P/C-G has an orbital period of 6.45 years and rotation of 12.4 hours. Prior to *Rosetta* reaching it, 67P/C-G had last passed through perihelion on 18 August 2002; the next passage was due on 13 August 2015, when it was hoped that the *Rosetta* orbiter and the *Philae* lander would have joined it for the ride.

67P/C-G was not actually the first choice of target comet for the *Rosetta* mission. Instead, it was comet 46P/Wirtanen. The reason for the change of plan was a two-year delay in the mission that was caused by the failure of an Ariane 5 rocket on the launch prior to *Rosetta*'s. A new target was needed and 67P/C-G was found to fit the window. The important factors were that the spacecraft had to be able to physically get to the chosen comet and the comet needed to pass close enough to the Sun, at the right time, to be active during its solar encounter. The scientists wanted to observe the comet as it interacted with the Sun, so for this reason it was important to find one that had not orbited close to the Sun too many times. As we've seen, every time a comet passes close to the Sun it is heated up, causing volatile material to escape the comet, taking dust and rock debris with it, shedding its outer layer like a reptile. The more times a comet visits the region close to the Sun, the less it will retain its original, and ancient, Solar System secrets, as they are moulted and left floating around in space.

A tale of two comets

With *Rosetta* successfully out of deep-space hibernation, it was ready to begin the phase of the mission that had been in the planning for over 20 years – entering orbit, mapping and landing on the surface of the comet. The scientists steered *Rosetta* onto a course that would allow it to catch up with comet 67P/C-G, approaching it from behind. However, before any of their dreams could be realised, the scientists desperately needed some accurate images of the space object they were about to get to know so well. Prior to close approach it was the astronomers, using ground-based telescopes and the Hubble space telescope, who had given *Rosetta* Mission Control in Darmstadt, Germany, the initial information about 67P/C-G. These studies estimated that the comet's shape resembled something like a dumpy potato, information that allowed the scientists to begin planning, and calculating, exactly how *Rosetta* would orbit the comet on arrival. Unfortunately, however, 67P/C-G had a few surprises up its sleeve.

As the spacecraft approached the comet it beamed the first – albeit slightly blurry and pixelated – images back to Earth from the OSIRIS (Optical, Spectroscopic and Infrared Remote Imaging System) camera and it soon became clear that the comet did not look quite like what the scientists were expecting from the telescope observations. As *Rosetta* got ever closer to 67P/C-G, and the images became a little less pixelated, the scientists started to see, to their utter surprise, that the comet looked more like two comets than one. The images Rosetta returned became clearer and clearer as the spacecraft homed in on its target and the shape of 67P/C-G soon became even more of a surprise. The comet was indeed just one object, but its shape resembled what can only be described as a rubber duck. The somewhat peculiarly shaped comet had a relatively small 'head', with dimensions 2.5km x 2.5km x 2.0km (1.5 x 1.5 x 1.2 miles), and a thin 'neck' region connected to a larger 'body', with dimensions 4.1km x 3.2km x 1.3km (2.5 x 2.0 x 0.8 miles).

The very irregular shape of the comet raised questions as to how it had formed. It was suspected that 67P/C-G either represented a contact binary comet – originally two comets that came together in a collision – or a single body that had experienced some loss of mass over time, such that it gradually developed a thin 'neck'. More investigations were required to work out which option was more likely, but the intriguing shape of 67P/C-G wasn't just causing a lot of excitement among the science teams, it was also capturing the public's interest. Nevertheless, the scientists who were calculating how *Rosetta* would orbit and land on the comet now had their work cut out. Determining how to orbit a perfectly round object is one thing, but this comet, with its two different lobes and associated lumps and bumps, required some careful thought to prevent the spacecraft accidentally colliding with it.

Rendezvous

In August 2014, *Rosetta* rendezvoused with 67P/C-G and from this stage onwards the level of detail in the images obtained of the comet's surface improved almost daily, helped by the fact that the spacecraft was within 100km (62 miles) of the comet nucleus. The scientists decided that *Rosetta* should first meet the comet when it was around 3AU from the Sun, because it was not expected to be very active at this stage. Of course, the orbiter was then to follow 67P/C-G as it approached the Sun, all the while carefully monitoring and tracking its increasing activity, one of the key research topics for the mission. However, the reality was somewhat different. Soon after *Rosetta* met 67P/C-G it captured some stunning images of jets of activity emanating from the neck region of the comet. Already scientists were learning that their understanding of the activity of comets as they approach the Sun is different to what they'd expected.

During September 2014, the *Rosetta* orbiter's work began in earnest when it entered its global mapping phase, powering itself around the comet to image its entire surface in detail. It made some very close approaches, being steered to within just 20km (12 miles) of the comet nucleus at times. High-resolution images were crucial to help the scientists assess surface conditions on the comet to minimise the risks when sending the lander down. Heading into a boulder-strewn field, or to the edge of a cometary cliff, would be avoided at all costs. In fact, the resolution of some of the images was so good that one pixel was only 50cm (19.6in) on the comet surface, which certainly helped, allowing the scientists to start ruling out landing regions that might be strewn with large boulders or crevices.

The aim was not only to image the entire comet surface but to also characterise the comet in general, determining its shape, rotation rate and orientation, gravity field, albedo (reflectivity), surface features and surface temperature. The *Rosetta* orbiter instruments such as RSI (Radio Science Investigation) and OSIRIS came into their own, measuring the volume of the comet to be 25km^3 (15 cubic miles). The VIRTIS (Visible and Infrared Thermal Imaging Spectrometer) instrument measured that 67P/C-G had a surface temperature of 205–230 Kelvin (-68–-43°C), and MIRO (Microwave Instrument for the *Rosetta* Orbiter) found the subsurface temperature to be just 30–160 Kelvin (-243–-113°C). VIRTIS also discovered that the surface of the comet was incredibly dark, porous and probably dry, something that wasn't very obvious from the grayscale images of 67P/C-G that the public had been so used to seeing. In fact, 67P/C-G is so dark that the scientists compared it to black toner ink, a rather unexpected colour for something apparently made of ice. The reason for this is that 67P/C-G's surface was not composed of ice, but instead of dust and organic matter, both dark with very low albedo.

Even before *Philae* had attempted its landing to begin investigating the comet, the *Rosetta* orbiter had begun sniffing the gases emanating from 67P/C-G. The ROSINA (*Rosetta* Orbiter Sensor for Ion and Neutral Analysis) orbiter instrument managed to detect several gases such as water, carbon monoxide, carbon dioxide, ammonia, methane and methanol. ROSINA also showed that the mix of gases streaming off the surface of the comet varied with location, with carbon monoxide being as abundant as water in some places, yet only 10 per cent as abundant in others. It was later realised that the comet was composed of different geological terrains, which fits nicely with the evidence seen in the images returned of the surface, even if they weren't fully understood at the time. These results indicated that 67P/C-G's distinctive regions were composed of inherently different materials. The implication is that they might have formed in different parts of the solar nebula, picking up different mixes of ingredients before coming together to form comet 67P/C-G. Interestingly, when the ROSINA data was compared with the results from other instrument teams, such as those of Ptolemy and COSAC (two of the lander instruments which we'll hear more about shortly), the variations in chemical signatures seemed to suggest that the two lobes of 67P/C-G could have had very separate histories. However, later analysis of these datasets instead indicated that 67P/C-G didn't tell a 'tale of two comets' after all. Its overall composition was found to actually be broadly homogeneous over the entire structure, despite local surface variations. This was an important conclusion to have reached as it indicated that the head and body of 67P/C-G were more likely to have formed in the same part of the early solar nebula, and at roughly the same time.

During the global mapping phase, the comet's activity was fairly low because it was still quite far from the Sun. However, the COSIMA (Cometary Secondary Ion Mass Analyser) instrument started collecting cometary dust grains that were already streaming off the comet, and continued to do this as it followed the comet closer to the Sun. The grains collected by COSIMA varied in size and texture which, on closer inspection, could either be described as compact and sturdy, or fragile and loosely bound. The latter actually shattered when they contacted the hard surface of the collector. However, the COSIMA scientists noted that in general the dust looked very similar to the comet dust regularly collected in the Earth's stratosphere, that of the IDPs, which also show a wide variation in textures.

Choosing somewhere to land

With the mission already progressing well, it was time for the *Rosetta* teams to focus on the main phase: choosing a landing spot for *Philae*. The decision was not so simple, because it was a bit of a contest between the scientists, who wanted to go to the most geologically interesting place, and the engineers, who were looking for the safest landing spot for their precious lander. Generally, the most geologically interesting location isn't the safest and so discussions were needed to find a compromise for the mission overall.

67P/C-G's many terrains were characterised by different textures. Smooth areas were more likely to make safer locations for *Philae* to land, but it was often the rougher, blockier regions that looked much more interesting scientifically. Active regions, such as those close to the location of jets of material exiting the comet, were also deemed too risky by the engineers. The potential for explosive eruptions of cometary gases, which were likely to become more frequent as the comet approached the Sun, was high. After a successful landing, the last thing the engineers and scientists wanted to see was little *Philae* destroyed and/or blasted back into space. However, the scientists who were trying to work out what the comet was made of knew that an active region would give the instruments plenty of cometary exhaust gases to sniff and analyse. A decision had to be made and there wasn't much time as a landing date had already been set and the days were ticking by.

At a meeting in Toulouse, France, in September 2014, 10 potential landing sites, initially named A–J, were whittled down to five (A, C, B, I, J), with Site J – later to be named Agilkia after an island on the River Nile – being chosen as the primary spot. Agilkia was located on the 'head' of the comet and site B, on the 'body' of the comet, was chosen as the back-up landing location. Both sites were viewed as 'scientifically interesting' with potential for activity nearby – but not too close – as the comet moved towards the Sun. The locations chosen contained some large boulders, which could pose a risk if the flight dynamic experts weren't spot on with *Philae*'s landing. There was no perfect landing site, with all of them being risky in their own individual ways, so a compromise had to be made. Throughout October 2014 the mapping of these sites continued, in order to characterise the chosen landing spots in even more detail, with the spacecraft getting as close as 9.8km (6 miles) from the cometary nucleus to improve image resolution.

Landing on a comet

If the *Rosetta* mission was a science-fiction movie, the viewer might have been hard-pushed to believe the rather crazy scenarios that were required to get the *Philae* lander to the comet's surface. The *Rosetta* orbiter had to release *Philae*, which had no power of its own, to land in a prescribed location on the comet. All the while, the comet was rotating, or tumbling through space. We must remember that such a tactic had never been attempted before, and despite all the images of the cometary surface obtained during the mapping phase, the scientists could not be 100 per cent sure from images alone what type of surface they would encounter – whether it be hard, soft, icy or composed of a deep layer of dust. These circumstances made the impending touchdown of *Philae* an anxious time.

67P/C-G is relatively small compared with other space objects, certainly much smaller than the Moon or other planets where scientists have some experience in landing spacecraft. There is little gravity on 67P/C-G and the fact that the comet wasn't spheroidal also complicated the landing calculations, because the gravity field around very irregularly shaped objects is much harder to predict, but luckily not impossible for the *Rosetta* scientists. Obviously, if the spacecraft jettisoned *Philae* when orbiting too far away from 67P/C-G, then there was a danger that it could just float off into space, not being attracted to the comet at all. Of course, with no power, *Philae* couldn't steer its course to the comet surface, so the release of the lander from the *Rosetta* orbiter required complex calculations and a careful spacecraft manoeuvre to place it in exactly the correct position at the right time and at exactly the correct speed. It was then hoped that *Philae* would fall to the chosen landing spot simply under the minimal force of 67P/C-G's gravity. The flight dynamic experts also had to take into account that the comet was spinning, meaning *Philae* would need to be released over the opposite side of the comet to the landing site, with the hope that it would gradually rotate into view by the time *Philae* reached the surface. If the scientists' calculations were incorrect, there was a high chance that *Philae* would either approach the comet too quickly, probably resulting in it bouncing back into space, or approach too slowly and completely miss the landing spot, or the entire comet. There was a lot riding on this one landing, and it was a tense time for all involved.

The landing date was set for 12 November 2014 and the countdown procedures began the night before. These involved a series of Go/NoGo checks to assess that all spacecraft systems were working correctly. The spacecraft was found to have just one major fault, which was *Philae*'s on-board nitrogen-cooled gas propulsion system. This was designed to be fired during the touchdown phase when *Philae* reached the comet's surface, and was intended to prevent the lander bouncing off the comet, and maybe back into space, should it experience a hard landing. Unfortunately, it was not possible to rectify this fault from Earth, so it was decided that the mission would have to continue regardless or they would miss the opportunity forever. The orbiter was manoeuvred to its final position prior to the release of its travel buddy, the *Philae* lander, ending up just 22.5km (14 miles) from the comet's centre. The landing sequence to release *Philae* began at 8.35 a.m. GMT on 12 November 2014 and *Philae*'s ballistic descent was imaged by the orbiter as the intrepid comet explorer began its seven-hour, 2.5km/h (1.5mph) fall through space. *Philae* unfurled its legs on the way until they were splayed like a three-legged cat ready for landing.

As is now well known, *Philae*'s landing did not have the outcome the scientists and engineers had hoped. *Philae* initially touched down precisely on target at Agilkia, confirming that the flight dynamics team had done a great job of aiming the little lander onto the comet. But very soon after this the lander took an unexpected turn, or bounce! Unfortunately, *Philae* encountered an impenetrably hard surface, later confirmed by the MUPUS (Multi-Purpose Sensors for Surface and Sub-surface Science) instrument when its touchdown data showed that the subsurface was as hard as solid ice and covered with a layer of dusty material. Without the all-important thruster to counteract any touchdown bounce, it had been anticipated that *Philae* would be lucky to achieve a safe arrival on such a surface without ricocheting off into space. *Philae* was also equipped with two powerful harpoons which were due to fire on touchdown, to anchor the lander to the surface to prevent it escaping the comet's weak gravity, but these failed to fire properly. However, without the thruster, even the firing action of the harpoons would have required a counter force to prevent *Philae* being flung off the comet before they had latched on.

There was another system available for the landing – *Philae*'s legs were designed to dampen the kinetic energy of the impact and the force of the landing was expected to drive ice hooks on each leg into the comet surface. These relied on the lander being on the surface for long enough at touchdown for them to work their way in, something that clearly didn't happen. Despite all of these landing aids, *Philae* encountered the hard comet surface and, without anything to hold it in place or push it back down, it rebounded and travelled back into space. This bounce took it up to 1km (0.6 miles) away from the comet until, luckily, 67P/C-G's weak gravity pulled it back towards the surface. A slightly stronger rebound caused by a faster approach could easily have sent *Philae* further off into space with no chance of recovery or return to 67P/C-G. *Philae* experienced another touchdown and bounce before finally coming to rest at a new location – a result of the continuing movement and rotation of the comet during the bounces. *Philae* ended up nowhere near the intended landing site at Agilkia. In fact, initially the scientists had no idea where it had ended up, although they did know it was still alive.

Philae's final touchdown site was later named Abydos, after one of the oldest cities in ancient Egypt. Thanks to images *Philae* beamed back to Earth of its final landing environment, the scientists discovered that it had ended up in a rather precarious position, tilted on its side and resting under a shadowy cliff. The good news was that one of the ice hooks appeared to have worked on the final touchdown, providing a small amount of stability.

Philae's unexpected landing situation was to cause the scientists a bit of a headache until they worked out exactly what had happened, what was still working and how they could proceed with the scientific experiments before the lander drained its primary power. At this stage, it was unknown how much sunlight *Philae*'s crucial solar panels would receive, which was the only way for it to continue operations after its primary batteries were empty. The clock was ticking but all was not lost because *Philae* had already been busy. In fact, it completed its primary science sequence, which was pre-programmed to begin automatically on touchdown as a series of predefined experiments lasting about 40 minutes. Of course, that meant it commenced this science sequence on the first of the three touchdowns, at the intended Agilkia site, and that it continued during the subsequent bounces. Obviously, the intention was that the lander would be stationary on the surface during this sequence, but the fact that it bounced didn't mean that measurements couldn't be made. Some of the first instruments fired up at initial touchdown were the Ptolemy instrument, a gas chromatograph mass spectrometer designed to measure light elements such as carbon, nitrogen and oxygen (the instrument discussed in Chapter 5 that has been adapted for many other uses on Earth), and COSAC (Cometary Sampling and Composition), a system designed to detect complex organic molecules and some of the same chemical species as Ptolemy. These instruments are described as having 'sniffed' the comet, opening their sample chambers to collect and analyse any gases that might be detectable at, or just above, the comet surface.

Sniffing the comet

Interestingly, the first set of results from COSAC and Ptolemy weren't exactly in agreement, despite them having collected their samples at the same time and from the same place. COSAC found an abundance of nitrogen-bearing species, which was an important result because nitrogen is an essential element for life and a fundamental part of amino acids. However, Ptolemy found very little nitrogen but lots of carbon dioxide, water vapour and carbon monoxide. It also found a small amount of carbon-bearing organic compounds, including formaldehyde, which is especially significant as it is implicated in the formation of ribose, a feature of molecules such as DNA.

The scientists were, at first, a bit confused by these results, as they knew both their instruments had worked as intended and that the results were well calibrated, so there began a serious stint of head-scratching to figure out why the instruments didn't agree with each other. After careful thought, it was found that both instruments were accurate, despite their different readings, because there had been a slight difference in the exact position of their sampling locations. The intake pipes for COSAC are located on the base of the *Philae* lander, and those for Ptolemy are located on the top. Even though the actual difference in height between these intakes is less than one metre (3ft), this corresponds to a real difference in the comet exhaust gases that were sampled by the instruments. COSAC is most likely to have ingested some of the volatiles in the ice-poor dust kicked up during the brief initial touchdown, whereas Ptolemy breathed in the cometary gases that existed higher above the surface, and the dust hadn't managed to rise high enough to be sampled by Ptolemy's intake. This was actually a fortunate and useful outcome. Even though the lander had spent only a few seconds in the sampling zone during its initial touchdown, its instruments still managed to pick up and sample different parts of the comet, both solid and gas. COSAC and Ptolemy were busy analysing these comet ingredients as they floated up above the comet's surface during *Philae*'s first bounce, long after they left the sampling area.

By the time the lander came to rest at Abydos, COSAC and Ptolemy had completed their first set of measurements and the orbiter had relayed many packets of data back to Earth via the *Rosetta* orbiter. The primary science sequence had been a success and this was great news for the mission teams, but *Philae* still had some remaining primary power and a new plan was hastily required for its remaining analytical time. *Philae* could operate for 60 hours on its primary batteries, after which it would need to switch to its main batteries that would be recharged by the solar panels. However, the race was on because the scientists suspected that the cometary cliff *Philae* had ended up next to would shade the lander's solar panels, giving them a just a fraction of the solar energy that had been hoped for – maybe 1.5 hours of sunlight a day as opposed to the six to seven hours they could have expected at the primary landing site, Agilkia. Sufficient solar energy would be vital for the continued operation of *Philae*'s on-board instruments, and without this energy the instruments wouldn't be able to start up, run and communicate their results back via the orbiter to Earth.

As *Philae*'s battery-recharging options were looking to be in jeopardy, the scientists knew that the little energy they had on the primary batteries shouldn't be squandered, so they came up with a revised plan for the remaining primary science sequence that could make the most of the energy available. The other problem was that it was unclear which of *Philae*'s instruments were in perfect working order. The lander had experienced three potentially damaging touchdowns on the comet surface, and it also appeared to be untethered and lying at an awkward angle at its final resting spot. To add to this, the scientists still didn't know where *Philae* was exactly. Because the *Rosetta* team couldn't get an aerial view of the precise location of the lander, and not knowing very well how the ground lay around it, they were loath to risk performing anything too risky, such as deploying *Philae*'s drill, in case it caused the unanchored lander to lift off the comet surface. This could even cause it to be launched off the comet altogether. However, there was always a possibility that the use of the drill to collect sub-surface samples, for example, might produce a counter-effect, knocking the lander over so it was the right way up. The decisions were hard to make and there were many to be made.

Despite all these problems, the good news was that the lander was on the surface of the comet with battery power and it had performed some science experiments. Despite the journey ahead, this part of the mission was already a huge success. Furthermore, the scientists had, unintentionally, achieved not just the first, but the first three comet landings in history because of *Philae*'s 'bunny-hop' across the surface of 67P/C-G. What followed for the scientists was an intense few days of planning meetings, changes to experimental plans, uploading of commands to the lander, and bundles and bundles of data being received back on Earth. All the while they were very aware of the ticking clock of battery power remaining. Many of the instruments got a second chance to perform experiments, meaning that analyses were made in two separate locations on the comet – the original landing site and the final resting place. This was an unexpected but handy outcome of the lander's bounce across the comet.

Towards the end of lander operations, the decision was made to deploy *Philae*'s drilling and distribution (SD2) subsystem, even though the scientists were unsure whether it would contact the comet surface because of *Philae*'s orientation. At this stage, with the main source of power trickling away, the scientists deemed that the use of the drill was worth the risk of it potentially moving the lander to a worse position, if it meant there was a possibility that the lander instruments would receive a solid sample for analysis. Plus, there was always the chance that *Philae* might end up in a better orientation or location where it was bathed in sunlight to charge its batteries. The drill was extended nearly 50cm (19in) from the lander before being retracted, but unfortunately it failed to collect any sample. This was a frustrating result considering that all the instruments appeared to be in good working order, but it had been worth a try. In the end, with *Philae* being placed into safe mode prior to its batteries giving out their last bit of power, it had fulfilled 80 per cent of its first science sequence in the 64 hours following its separation from *Rosetta*. The bonus was that it had collected these data in more than one location, allowing scientists to see more of the comet than they'd originally thought possible.

***Philae*'s final gasp**

Luckily, the plucky little *Philae* lander wasn't ready to say its final goodbye just yet; after a journey of 10 years it wasn't going to give up without a final gasp. In June 2015, just two months before perihelion, *Philae* had received enough sunlight during seven months of hibernation on the comet's surface to wake up once again and contact the orbiter. With continued sunlight for the foreseeable future as 67P/C-G approached the Sun, the scientists hoped that continuous lander operations might be possible. The only issue was that, because the scientists still didn't know the exact location of *Philae*, it was tricky for them to plan an orbit for *Rosetta* to be in visibility of the lander to communicate with it, to upload commands and pick up any data *Philae* might have stored. The *Rosetta* teams were not keen to bring the orbiter too close to the comet because 67P/C-G was fairly active by this stage. It was throwing off a great deal of dust, sometimes resulting in explosive jets of cometary material exiting the comet. Not only did the scientists not want to risk damaging the orbiter with the impact of a fragment of high-velocity dust, but they also didn't want to risk flying *Rosetta* into a cloud of dusty debris that could potentially confuse its 'star-tracking' navigation system. Sun glinting off a piece of dust close to the orbiter can look uncannily like a far-off star. *Rosetta* uses the stars to figure out where it is located and if it mistakes an errant piece of comet dust for a star then it will be utterly confused as to its location and could easily wander off into space or, worse, crash into the comet.

At this point, the orbiter was still working perfectly and had a lot of science ahead of it as the comet approached the Sun and came out the other side, whereas the lander was a somewhat riskier proposition. It was decided that bringing the orbiter too close to the comet was just too much of a risk for the overall mission, especially as there was no guarantee that it would find the lander, which itself might not have a good enough source of continued power to be able to perform any experiments anyway. By 9 July 2015 the lander had made seven intermittent contacts with *Rosetta*, but the communication links that were established were too short and unstable to allow for any commands to be uploaded and that was the last the scientists heard from the intrepid little lander. *Philae* was, once again and for the final time, asleep on the comet, clinging on for dear life but having the ride of its life.

In a heart-warming turn of events just a month before the mission was set to end, *Philae* was spotted by the OSIRIS camera as the orbiter came within 2.7km (1.6 miles) of the comet surface, allowing images with a resolution of 5cm/pixel. The body of *Philae*, with two of its three legs sticking out, was found wedged in a shadowy crack on the side of 67P/C-G, which explains why it had so much trouble communicating with the orbiter. The science that had been completed could now be placed in the wider geological context of the region, helping to make a fitting end for the successful little lander.

Up close and personal with 67P/C-G

It may sound like *Rosetta* was beset with problems, but the mission team never claimed that landing on a comet would be easy. Despite their confidence that the mission could be a success after all their years of hard work, the fact was that they were setting out to achieve things that had never been attempted before. Even with the best planning they could do, and their control of every known factor, landing a spacecraft on a speeding comet was a huge feat of scientific and engineering magic. The fact that the scientists then managed to get the lander to perform scientific sequences, and return data to Earth, confirmed the *Rosetta* mission as an undeniable success. Much of the science carried out by the lander complemented the orbiter data, giving the scientists even more confidence in their results.

The COSIMA instrument on the orbiter collected dust throughout the mission and found it to be dry, without ice; it was suggested to have originated from the dusty outer surface of the comet that gave 67P/C-G its dark colour. In fact, the MUPUS instrument that was located on the *Philae* lander, which used a hammer to pound the surface of Abydos, confirmed this to be the case. The comet was covered with a layer of dust 3–10cm (1–4in) thick, overlying a much harder compacted material. Again, this seems a bit counter-intuitive. Comets are meant to contain ice, they are known as dirty snowballs after all, so it is a bit confusing that no ice can be seen on 67P/C-G's outer surface. However, it turns out that it's probably *just* the outer surface of the comet that is ice-free, as all the frozen volatiles are protected below it, forming the solid comet interior.

The explanation for this structure is thought to be related to how the comet interacts with the Sun each time it passes. The current external dust layer is probably a relatively recent feature formed at the previous perihelion as the frozen gases present in the comet near to the surface were heated up and sublimated. This left behind an outer crust of dry dust and organics. As the comet comes back towards the Sun on its next orbit after its jaunt through deep space, the dusty outer shell is thought to be shed as the frozen volatiles sitting below are, once again, heated up and start to leave the comet. As the volatiles sublime away they take with them the dry dusty outer layer that's in their way. Of course, this is some of the dust that produces the comet's coma and dust tail, and represents the type of dust collected by COSIMA.

This cyclical process exposes a new dust layer during perihelion every six years as the outer layer of the comet is once again heated and dehydrated. The result is that the comet gradually loses mass with each orbit of the Sun and it gets smaller and smaller over time. Estimates suggest that 67P/C-G loses a surface layer of material around 1m (40in) thick during each approach to the Sun. It is understandable that, over time, an active comet can become 'extinct' as it uses up all of its volatiles. It can then no longer exhibit the classic cometary behaviour of a beautiful coma and tail since it is the volatiles that drive this behaviour: once they are depleted the comet is no longer active.

COSIMA also discovered another interesting feature: the type of dust grains collected changed as the comet got closer to the Sun. At first, when the comet was 3.5AU away, and up until around 2.3AU, the dust collected was dominated by the fluffier grains, representing the very outer fragile dusty layer of the comet. However, as the comet passed through 2AU, and even closer to the Sun when cometary activity was increasing and becoming more violent, it resulted in explosive jets of activity at the comet surface. At this stage COSIMA collected more of the compact, dense type particles. These are likely to be heavier and require more energy to lift them from the surface, or they emanate from the interior of the comet, explaining why they weren't seen until the comet was more explosively active.

The evolution of comets, and their dust production during perihelion, was not understood very well before *Rosetta*. Observing how a comet interacts with the Sun is important for our understanding of how comets evolve and change with time as they periodically enter the inner Solar System. This allows scientists to better predict their future behaviour, which might be very important if one is found to be headed our way and we want to do something to stop it impacting Earth. In fact, understanding 67P/C-G's dust was so important that COSIMA was just one of three instruments on the orbiter that was designed to measure it. It was joined by GIADA (Grain Impact Analyser and Dust Accumulator), which was able to measure how many dust particles came off the comet, how big they were and how fast they were moving, as well as understanding their composition based on their density. In fact, the GIADA instrument results support those from COSIMA very well to confirm that there is a difference in density between the compact and the fluffy particles. GIADA scientists suggested that the compact particles might represent dust that has been processed in the inner Solar System, whereas the fluffy ones are more likely to represent the primitive 'unprocessed' dust, material that may pre-date the

Solar System. Such an interpretation is made easier in light of the *Stardust* mission findings (discussed in Chapter 7).

Then there is also the MIDAS (Micro-Imaging Dust Analysis System) instrument, designed to analyse the very smallest particles, those around 1 micrometre across. MIDAS is particularly adept at performing 3D scans of these tiny particles – so small that even a human hair would be too large for it to analyse. The reason it focuses on the tiniest of particles is that these are the dust grains that are essentially invisible to telescopes because of their small size, so capturing them in space is the only way to study them in detail. The 3D scans allowed the scientists to study the structure of these particles and, therefore, determine whether they are made up of even tinier particles, possibly interstellar dust grains, or if they are a single grain.

Such information is important to help scientists understand how, and even where, the comet dust grains formed and, ultimately, how the comet came together. A higher proportion of the compact, single grains of dust would show that the comet collected up more of its dust from the inner Solar System, indicating that it might have formed closer to the Sun. A higher proportion of inner Solar System dust may make the comet seem more 'asteroid-like', but we know, particularly from the *Stardust* mission findings, that inner Solar System dust can be expected in comets. Something that is not yet known is how much of this 'inner Solar System' dust should be expected in a 'normal' comet, if we can label any comet as 'normal'. To help with this problem it was useful to understand the internal structure of 67P/C-G, and to work out its overall density. Knowing the density of individual particles, even if they could measure every single one, doesn't allow scientists to work out the overall density of the comet. The problem is that they need to know how well the material that comprises the comet is compacted together. If the comet is not very well compacted, containing lots of empty cavities, then it would be expected to have a lower overall density.

It was calculated that 67P/C-G has a density of 0.53g/cm^3, which is just half that of water (1g/cm^3). Since even ice is about 0.9 g/cm^3, such a low overall density for the comet is a slightly surprising result given that we know it contains rock, evidenced by the dust collected by the orbiter, which has a higher density than water. To account for this, the scientists concluded that the rocky and icy materials that form 67P/C-G must be packed together very, very loosely and that the comet can be described as fluffy, or spongy. Alternatively, the comet could be composed of large blocks of dense compact dust separated by big gaping holes that would act to lower the overall density. However, by using radio waves to investigate the interior of the comet, the scientists managed to conclude that the latter scenario couldn't be accurate.

The way this worked is that the *Rosetta* orbiter sent radio waves back to Earth, communicating with it, as it journeyed around the comet. Despite 67P/C-G's relatively small size, it still has gravity associated with it that acts to tug lightly on the orbiter. If the internal structure of 67P/C-G contained big empty spaces, then these would affect the force of the gravity exerted on *Rosetta* as it orbited the comet. Even if the smaller force exerted on *Rosetta* by such voids is only very slightly different, the comet would still tug less on the spacecraft at various times during its orbit, as it passed these voids, resulting in a very delicate change in the orbiter's velocity. This change would result in a measurable difference in the frequency of the radio waves received on Earth that were sent by *Rosetta*. This may sound a bit complicated, but it is actually very simple. If the radio waves received on Earth showed a range of different frequencies, then the scientists could conclude that the comet had a complicated structure containing void spaces that acted to alter the gravitational effect on the orbiter, apparent as a change in its velocity. However, the scientists found the opposite to be true: *Rosetta*'s velocity stayed the same as it orbited around the comet, suggesting that 67P/C-G must have a homogenous structure down to a very small scale, probably of the order of centimetres (*i.e.* it doesn't contain big gaping holes). Therefore, the scientists could conclude that 67P/C-G is 'fluffy' in structure to give it its overall low density.

These results came from the use of RSI (Radio Science Investigation), which correlated with those from the CONSERT (Comet Nucleus Sounding Experiment by Radio-wave Transmission) instrument that had earlier studied the small lobe of 67P/C-G – the duck's head – by passing radio waves through the comet between the lander on the surface and the orbiter out of sight on the other side. The CONSERT scientists concluded early on that the duck's head was very loosely compacted, with a porosity around 75–80 per cent, but with a structure that was fairly homogenous at the scale of tens of metres – so containing no big holes. It was certainly good to see that these different experiments agreed with each other and it gave the scientists further confidence in their conclusions. The other great thing about knowing the density and overall structure of the comet is that it allowed scientists to calculate how much it weighs, with 67P/C-G coming in at a whopping 10 billion tonnes!

The comet's ice

There has, so far, been a lot of focus here on the rock dust that makes up comet 67P/C-G, mostly because it is so useful in helping to understand where the comet formed and under what conditions. But the ice within 67P/C-G also had an interesting story to tell. The *Rosetta* scientists found that the dust-to-ice ratio of the comet was fairly high, meaning that it was very dusty – an icy dust ball, shall we say. In a study carried out by the ROSINA instrument, it was possible to estimate the structure of the water ice in the comet, which could have been either crystalline or amorphous. These terms refer to the geometric arrangement of the molecules in the ice. Crystalline ice is the type you'll be familiar with: it forms all the ice found in the natural environment such as in snow and polar ice caps, as well as the type in your freezer. The molecules are arranged in neat, repeating, geometric patterns. The molecules in amorphous ice, on the other hand, are arranged in a disordered state with no regularity. These distinctive configurations of ice form under different conditions. So, having some idea of the structure of the ice in 67P/C-G meant the scientists could speculate on where the comet formed.

Amorphous ices are thought to be 'pristine', being more likely to have formed in the calm and cold outer reaches of the solar nebula, or in the interstellar medium. Amorphous ice is very common in space and when it forms, it efficiently traps large amounts of volatiles in its highly porous, yet disordered, structure. This is in stark contrast to crystalline ice, one type of which is known as ice clathrate. This is a structure that physically resembles ice as we know it, but forms a distinct type of lattice that locks different molecules – volatiles such as methane for example – into its cage-like structure. Clathrates are thought to have formed in the region of the cooling nebula located closer to the Sun, an area which experienced slightly higher temperatures (albeit temperatures that are still cold enough to form ice) and more turbulent conditions. As such, as ice sublimates from the comet, it is possible to detect whether it was once amorphous or crystalline in structure by the way it releases its volatiles. These distinctive patterns of volatile release can be picked up in the data collected by the *Rosetta* instruments. The patterns observed in the ROSINA data collected from the comet were not indicative of the sublimation of amorphous ices but, instead, appeared to show that 67P/C-G could contain clathrates. An abundance of clathrates in 67P/C-G would indicate that it, and possibly other Jupiter-family comets, formed closer to the Sun than is generally appreciated; otherwise they would be expected to contain 'outer solar nebula' amorphous ice. This is yet another intriguing result from the *Rosetta* mission that will be useful for many scientists working on understanding the formation environment and location of comets in the early solar nebula, even if these preliminary findings need further review.

The challenge now for scientists studying 67P/C-G is in combining all these interesting findings to build up a picture of the comet's history. While it certainly contains an abundance of fluffy dust, indicating formation in the outer disc, it also contains dense, sturdy particles that are expected to have come from the inner Solar System. While there is evidence to suggest that the comet's ices might be present in clathrate form, which would indicate that 67P/C-G's ice is not outer disc in origin, there is also evidence from the ROSINA instrument that the deuterium-to-hydrogen ratio (D/H) of the water in 67P/C-G is high. This is something that is indicative, as we currently understand it, of water originating in the outer solar nebula. Interestingly, the water in 67P/C-G is found to have a higher D/H than other Jupiter-family comets, but caution should be applied here, because not that many have been measured to date and they have also been measured by different methods that might not be comparable. Nevertheless, such a result could just support the idea that the Jupiter-family comets formed over a wide range of heliocentric distances, or at slightly different times, with the lower D/H ones forming closer to the Sun, and/or later, than 67P/C-G. What would be very interesting to know is how these D/H ratios vary with the structure of the ice in the same comet. Presumably we'd expect the comets with lower D/H to contain water ice in a clathrate structure rather than it being amorphous, but perhaps the comets have yet more surprises for us.

While these lines of evidence seem to be at odds with each other, with some suggesting the comet formed in the outer disc and others suggesting it formed a bit closer to the Sun, they will clearly be useful in helping to piece together the history of how the comet came together, and where.

Even though the *Rosetta* mission has provided a huge amount of data, it has only looked at one comet of the many billions that are out there. These titbits of data provide clues that help scientists to put together the different pieces of the early solar nebula jigsaw, but it can seem like frustratingly slow progress. Yet, many of the findings from the *Rosetta* mission parallel those from *Stardust*, the only other mission to get up close and personal with a comet. In fact, the simplest of the amino acids, glycine, was also detected on 67P/C-G, as it had been at Wild2. *Rosetta* further showed that the glycine was closely associated with the dust at 67P/C-G, additional evidence that organic matter can act as a glue to hold together the icy, rocky, fluffy dust particles that make up comets, and something that has long been predicted from the study of IDPs. As we can see, *Rosetta* highlights the importance of space missions in our understanding of the formation of these enigmatic ancient, icy objects and shows us that future missions, whether they be to other comets, asteroids or planets, can build on these findings to answer the key questions we have about our Solar System.

Mission end

Space missions eventually have to come to an end in one way or another. Whether they fizzle out quietly as their precious power runs out, or they come back to Earth with their haul of space goodies, it is always a time of mixed emotions for the scientists involved. After *Rosetta*'s encounter with 67P/C-G, the scientists knew that the mission couldn't keep going for eternity. The little lander could be forever sat on the side of the comet, wedged in a little shadowy crack, but it was to remain in deep-space slumber. However, the orbiter was still alive and tracking 67P/C-G through space. The only problem was that the comet was, once again, on course to enter deep space where communications would be virtually impossible, not forgetting that *Rosetta*'s power wouldn't last forever.

The scientists had a few options open to them: they could simply turn off the communications and wait to see if the spacecraft would reawaken during the next perihelion; they could bid the *Rosetta* mission farewell and leave it orbiting the comet while heading off into deep space; or they could be more daring and try a final bit of science. They chose to be adventurous, something that involved ending the mission in a blaze of glory. Well, to put it more accurately, to undertake a controlled descent of the orbiter onto the comet surface – effectively a crash landing, counting as their fourth comet touchdown. In September 2016, the *Rosetta* orbiter began its heroic descent to meet 67P/C-G, while capturing high-resolution images and studying the comet's gas, dust and plasma environment on the way down. *Rosetta* completed its final mission while beaming back data up to the last moment of impact. It was an emotional way for the *Rosetta* teams to say goodbye to such an amazingly hard-working, productive little spacecraft and it seemed fitting that *Rosetta* could end its mission as the impressive workhorse that it always was.

CHAPTER 9

Space Mining

Mining in space may sound like science fiction, or the kind of pseudo-scientific nonsense that can often feature in blockbuster movies. Yet, it turns out there are tangible plans in place to make space mining a reality. Incredibly, these proposals to mine space for its resources could come to fruition in the next few decades. Building upon the recent successes of space missions that have rendezvoused with comets and asteroids, sampling, analysing and even landing on them, several visionary entrepreneurs have taken it upon themselves to establish space-mining companies. The aim of these endeavours is to repeat in space what our Victorian forebears did around the world – generate huge profits from mining natural resources. Such companies have a shared motivation: to make space mining profitable while also helping to open the 'final frontier' into space beyond our local neighbourhood. Certainly, some of the technology is already available to take the initial steps towards modest space-mining efforts, but there is still a long way to go to make these ambitious proposals a reality. Despite the huge investment that has already been made, there is still the issue of whether space mining can ever make viable economic sense. Importantly, mining in space has the potential to pay for itself, making future space exploration less of a drain on taxpayers. It might provide support for 'blue skies' research; indeed, the future of space science research may rely on commercial space mining for funding, with valuable trade-offs between private companies, government agencies and academia.

Experience shows us that it costs hundreds of millions of dollars to develop and launch just a single spacecraft. The cost of developing the technology and expertise for a space-mining programme can easily run into trillions of dollars, so it is hard to understand how commercial space companies can possibly reap enough rewards from space rocks to make it worth all the effort. However, the fact that the metal deposits alone within a single asteroid could be worth tens of trillions of dollars means that, done correctly, there is potential to generate a profit from space, even with the colossal initial set-up costs. Space contains every element known to humankind, in virtually infinite amounts – the sky is literally not the limit. The gains from space-mining efforts won't just be financial, either. Through collaboration with government space agencies, science will gain, too, providing valuable knowledge about nearby space objects.

Proximity and purpose

Our closest extraterrestrial neighbour is the Moon, and it is likely to play an important role in future space-mining activities, particularly with the announcement of the Google Lunar XPRIZE that promises $25 million to the first private company to land on the Moon, travel at least 500m (0.3 miles) and send back high-definition images. Moon Express is ahead of the pack in this respect, having become the first private company to obtain permission from the Federal Aviation Authority (FAA – the US national authority that regulates civil aviation) to land a spacecraft on the lunar surface.

Water and helium are in high abundance on the Moon, both of which could be useful in space. At the time of writing, there are companies looking at mining water from the Moon's poles to extract hydrogen and oxygen to use as rocket fuel, but the Moon may also contain a rich inventory of rare metals. The problem is that the lunar resources are in limited supply. For example, helium may only be found in the outer layers of the Moon's surface, meaning that vast swathes of the Moon would need to be dug up and processed to obtain a constant and economic supply. Although there could be many years of helium availability, the question is: are we happy to dig up the surface of the Moon, changing forever the aesthetic of this iconic object which is clearly visible from Earth? Individual asteroids and comets may contain a more limited supply of resources, but, once the techniques have been developed to prospect and mine on a single small object, they could be applied, repeatedly, to all the other millions of asteroids and comets available out there – a seemingly infinite number. Additionally, if we want to develop a base on Mars, and explore further into deep space, we will need to understand how to utilise smaller bodies.

Although mining on the Moon is clearly an exciting prospect – and we won't completely ignore it – in this chapter we will have to contain ourselves a bit and focus on the smaller, and more distant, space objects that are the emphasis of this book. Some of the main targets for initial space-mining efforts are asteroids, but comets may also be of interest. Asteroids and comets represent a potentially rich resource of precious metals and, in many cases, water. The high cost, and limited supply, of many metals on Earth makes them an obvious focus for mining efforts for anyone looking to turn a profit. However, figuring out how to efficiently mine water might be just as lucrative. The volume of water required for long-duration space missions, particularly any future missions involving humans, would result in very high launch costs from Earth – it takes a lot of energy to escape Earth's gravity well. Water isn't just needed for sustaining human life, but may also be useful as fuel. Future exploration of space by humans will almost certainly involve having to locate a water supply away from Earth, simply because it won't be possible to economically, or physically, take enough on board prior to launch.

Not only can asteroids and comets contain precious metals and water in high abundance, but these small objects can also be relatively easy to get to, with some being energetically closer than the Moon, requiring less rocket thrust to reach them.

The relatively small size of comets and asteroids on planetary scales also gives them a low surface gravity which makes them easier to leave than the Moon, requiring less energy to blast back off their surfaces with a spacecraft laden with mined goodies. In addition, straightforward surface contact with some asteroids and comets may be enough to break them up, simply because they aren't very well consolidated. In some cases, it might be that drills and excavators are not even required. However, at the same time, the lack of a strong gravitational pull to these objects presents many challenges for space mining. Spacecraft and mining equipment would need to be tethered to the surface of small objects to stop them floating off into space, which presents some obvious challenges if drilling is required. This shouldn't be impossible to achieve, though – space agency missions have already had to deal with this problem. Such technologies are not particularly complicated, but what *is* complicated is getting the technology onto the surface of the space object successfully in the first place; this is the part that needs practice and patience. The lack of gravity on small space objects can also pose serious risks to orbiting spacecraft. If a lot of surface material is kicked up when landing the spacecraft or carrying out mining activities, then it may collide with orbiting aspects of the mission.

Where to process the space goodies?

It seems likely that landing on, and drilling into, small space rocks will be made possible by space-mining companies and space agencies within the next few decades. After all, the *Rosetta* mission has already achieved nearly all of these goals using relatively dated technologies developed in the 1990s. In addition, the *Hayabusa* and *OSIRIS-REx* missions have tested, or will test, similarly useful technologies. However, even if space-mining companies achieve their initial goals, it will be just the first step in a long process of events that will unfold before profit can be made from these rocks. Even if resources can be extracted from small bodies in space, the next step will be in processing the mined resources, figuring out how, and where, this should take place, and who owns the rights to them. Processing of mined raw materials could either be done on the space object itself or the resources could be moved to a different location for processing. If the processed materials are required for use in space – maybe for fuel or to build infrastructure – then it wouldn't make much sense to return them to Earth for processing, due to the cost of relaunching them back into space. It would, therefore, be more efficient and cost-effective for space-mining companies to process raw materials in space.

The challenges for processing mined resources in space are many, and they shouldn't be underestimated. Processing mined products in a low-gravity environment is something that could require very different techniques to those used on Earth. For example, the use of simple gravity separation, something which is taken for granted during terrestrial mining, cannot be relied upon in space, particularly on small bodies such as comets and asteroids. Locating a processing plant in low Earth orbit, or on the Moon, where the gravity is just a sixth of that on Earth but considerably more than on asteroids or comets, would still require different techniques to those used on Earth. Nevertheless, the low gravity provided by the Moon is better than the near-zero gravity on a small comet or asteroid. However, the lunar gravity would still mean that the cost of launching a spacecraft laden with heavy mined products off its surface would be high, significantly eating into space-mining profits. The Moon does have some other advantages, though. For one, it stays put. The Moon remains, for practical purposes, the same distance from Earth, which makes planning repeat missions much easier, particularly in relation to the NEAs and NECs that are whizzing in and back out of the inner Solar System at great speed. This means that a lunar-based processing plant would be easier to maintain, update and fix, with the possibility that humans could one day live and work there.

The idea of sending humans to work on smaller space objects such as comets and asteroids is certainly exciting but is something that will be significantly harder to achieve, despite what you might have seen in the movies! Sending crewed missions to the Moon, or establishing a stable orbit around it, could mean that the entire mining process needn't necessarily be automated. Even if it was, humans would still be required for trouble-shooting the machines when they inevitably developed faults. Despite great progress in robotic technologies in recent decades, the trouble-shooting capabilities of the human mind and dexterity of the human hand cannot be beaten, except perhaps in simple and repetitive tasks. In this respect, having humans in place would be much more helpful than having them work remotely from Earth, thus making the Moon an attractive prospect.

Even if the Moon isn't utilised for the development of a space infrastructure base for full processing of mined materials from smaller space objects, then it still offers advantages because of its proximity to Earth and stronger gravity in relation to asteroids and comets. This may mean that it can act as a testing ground for mining technologies that could eventually be sent to asteroids or comets.

Which space objects are going to be exploited first?

Initial prospecting in the Solar System will be a balance between the type of resource sought after and the ease of reaching the target. As we've already seen, there are many different types of asteroids and comets, which are composed of varying complex mixes of rock, metal and ices. The C-type asteroids contain lots of water, much like the comets, whereas the M-type asteroids, being the segregated core of differentiated asteroids that have lost their rocky mantle, have high concentrations of metals. S-type asteroids, on the other hand, are a mixture of rock and metals. The important thing is that the asteroids can contain very high abundances of some important ingredients, in particular the precious metals, which are found in limited quantities on Earth but which we rely on in a wide range of technologies and industries.

If materials can be efficiently mined from asteroids, and returned to Earth, then they can provide us with more metal than we'd ever be able to collect from the Earth's crust. The impetus for space mining is not just about mining materials to return to Earth, though, because these space rock ingredients may be useful in helping us to further explore the Solar System. For example, mined metals could be used to enable 3D printing in space. Human space missions, to date, have relied on taking everything that they think will be required with them at launch as, so far, there has been no capability to fabricate or manufacture items in space. In the future, this may not be necessary, as space can provide all the ingredients needed to survive, and thrive, away from Earth. Plus, space can even provide us with fuel, a key requirement for long-duration missions, reducing the need to transport fuel from Earth, or having to rely on solar energy which decreases with distance from the Sun.

The most obvious place to focus initial space-mining efforts is on the small bodies such as the NEOs that orbit the Sun within 1.3AU. These objects are of high interest because they represent an almost limitless supply of resources in space, containing tens of trillions of tonnes of material within 100 lunar distances of Earth, just a small step in space terms. As we've seen, tens of thousands of NEOs are known today, with that number increasing daily, and there could be more than one million of them. However, fewer than 1,000 observed NEAs, the most common type, are more than 1km (0.62 miles) in diameter. NEAs are seen as 'easy pickings' as they are relatively simple space objects to get to, compared with, say, Mars. However, their small size and lack of significant gravity fields also present some obvious challenges for mining activities.

The orbits of the NEOs vary by millions to hundreds of millions of kilometres, so the distance to get to them from Earth depends on how far through their orbital journey they are. If you want to send a spacecraft to meet an NEO, then you might want to approach it as it's re-entering the inner Solar System so that you aren't having to chase it as it's getting further and further away from Earth. Entering deep space presents challenges for many reasons, the most important of which is the great distance from the Earth and Sun, making it much colder, with little solar energy and very long communication times. However, chasing down NEOs requires waiting for a suitable launch window that will provide the quickest and least expensive mission cost, but which might still involve planetary fly-bys to gain gravity assists. Suitable launch windows are dependent on the individual object's orbital period and they tend to occur infrequently, perhaps every few years at best. If you manage to get your spacecraft to one of the more easily approachable objects in a reasonable timeframe then, even when the spacecraft has intercepted its chosen object, the next problem is keeping it in formation during its continued orbital journey around the Sun. During this time the space object, and accompanying spacecraft, will be on a periodical plunge into deep space at aphelion, when the object is at its farthest point from the Sun, a time when it might be deep in the heart of the asteroid belt. Of course, there is also the opposite problem when the object is at perihelion, its closest approach to the Sun when, even if it doesn't perform a very close approach, it could experience dangerous and destructive high temperatures. Both the spacecraft and the asteroid or comet could be blown to pieces by the force and heat of the Sun at perihelion, as happened to Comet ISON in 2011, an event which scientists were unable to predict with any certainty. Of course, even if the object stays intact during perihelion, it will still become more active as it's heated up near to the Sun. At this stage, it would be appropriate for the accompanying spacecraft to move back to avoid being pummelled by escaping comet or asteroid debris as the object sheds layers of dust and rocks.

As mentioned, the relatively small size of the objects of interest for space mining means they have little gravity, which can be a blessing and a curse. The spacecraft won't be able to enter a true orbit, since it won't be gravitationally bound to it. This makes mission planning more complicated and costly, since fuel will be needed to keep the spacecraft in formation with the object. At the very least, orbiting at close proximity will require a well-constructed and thought-out shape model of the object of interest to allow for extremely accurate mathematical orbital planning to be undertaken to avoid an accidental collision. You will recall, this was exactly the case for the *Rosetta* mission. Of course, a shape model would be required, in any case, to plan mining activities so it's likely this wouldn't add a significant amount of extra workload to the plan. However, a lack of gravity makes it relatively cheap and easy for the lander to depart the object laden with its cargo.

In 2013, a group of 12 NEAs were identified as potential candidates for space mining, being seen as objects that could be mined with present-day rocket technology. These were defined as 'easily retrievable objects' (EROs), chosen because they were not on orbits with too high an orbital inclination or that were too eccentric. This means that they were not in the same plane as Earth's orbit but close to it, and they were relatively easy to reach, requiring only small changes in velocity in space to catch up with them. One of the other desirable qualities of EROs is that it would be possible to move them to a more suitable orbit relative to Earth. Yes, literally physically move them from their orbit! This shift might put them into another stable orbit, around the Moon or the Earth, or to a so-called Lagrange point: a gravitational sweet spot in space where the combined gravitational forces of two large bodies – such as the Earth and the Sun or the Earth and the Moon – are balanced. There are some Lagrange points located within a few million kilometres of Earth, making them relatively easy to get to, and if an object, such as a captured asteroid, is placed in one of these locations then it can effectively 'hover' there for free. This also means that any spacecraft approaching it in the future, for prospecting or mining activities, can get there without expending too much energy and will be able to join the object on a stable orbit that won't require energy to maintain.

Nevertheless, at this stage, not very much is known about the EROs that have been identified, apart from the fact that they are quite small, all being less than 100m (328ft) across, because they've only been glimpsed remotely from Earth. Their exact composition remains a bit of a mystery and so it might be that some, or all, of them don't represent suitable targets for mining. The good thing about the EROs is that their small size makes them safer to move, particularly if something were to go wrong during the redirection process. If the object were accidentally placed on an Earth-crossing orbit, then its small size would almost certainly result in it burning up on atmospheric entry to Earth so that it shouldn't pose a risk to us. In this sense, these small objects can work as a logistical stepping stone to mining much larger and more lucrative asteroids, as they allow many of the main techniques to be tested relatively safely.

How do we begin looking for resources in space?

Even on Earth, extensive exploration and prospecting are necessary before mining of raw materials can begin in a new region. It is no different in space. Any asteroid, comet or planet of potential interest for mining will need to be surveyed, assessed and analysed in detail, to quantify its overall composition and size, to work out if it represents a worthwhile, and economical, supply of resources in the form of metals or water. How the resource of interest is distributed throughout the object would be a crucial consideration. If we think about two separate asteroids that have a similar bulk composition, they could contain the same quantity of metal overall, but in one the metal could exist as one large, amalgamated lump while in the other it might be mixed in thoroughly with other rocks, possibly at a small scale. This would determine how easy it would be to physically remove the metal from the object, possibly requiring vastly different mining and processing techniques. Other key features that would need to be assessed would be how well consolidated the object was and whether it had a firm surface – features that would affect the type of drilling equipment required to break it up, and whether such equipment had a suitable surface on which to anchor itself.

Historically, exploration and prospecting for raw materials on Earth have been rather humble yet time intensive, often being carried out by humans on foot, endlessly searching stream beds for shiny rocks that might reveal the presence of valuable minerals upstream. Space mining will be a totally different story. We won't be packing astronauts off on long, arduous journeys in the hope of finding a speck of a precious metal. Instead, we'll be operating automated or remotely controlled systems far from Earth from the outset, without the direct input of humans (who will be safely based on Earth), millions of kilometres from the objects their spacecraft are studying. However, despite the need for complex technology in space-mining activities, it's all relative. We've been exploring space using robots for several decades now, and have developed a multitude of complicated and sophisticated machinery. In comparison, the sort of technology that will initially be required for prospecting space objects for potential mining activities needn't be too complex. The primary prospecting steps can easily be carried out using Earth-based telescopes, or relatively simple telescopes in low Earth orbit. Additionally, specialist satellites can be launched with a precise job in mind that would initially be as basic as locating potential space objects, estimating their size, shape, density and exact orbit, as well as whether they are rocky or metallic, or contain an abundance of volatiles. This would be a relatively simple place to start with current technology.

Once a space object is located with potentially suitable composition for mining, it would need to be mapped in much greater detail to establish the exact composition of metals, rock and volatiles, and how they are distributed. This would allow for an assessment of whether the object represents an economic resource of materials. However, because many of the objects of interest are on the small side, NECs and NEAs that are tens of metres in diameter, assessing them in detail from Earth, or even from telescopes in low Earth orbit, is not currently possible. Obtaining high-quality geochemical and geophysical data with a high enough spatial resolution to determine the exact composition and location of different materials would almost certainly need to be done in space itself, using orbiting spacecraft equipped with more advanced scientific instruments. At this stage, it would also be important to obtain rock samples from the surface of the objects for detailed chemical characterisation. A lander probe may be necessary for this, but to avoid having to perform a potentially risky landing on an object with unknown surface conditions, it might be better to use a spacecraft that could grab a sample from the surface in a touch-and-go style approach such as that employed by the *Hayabusa* and *OSIRIS-REx* spacecrafts. Such a spacecraft could even sample multiple targets in a single mission. The rock samples collected could be analysed by equipment on-board or be returned to Earth for analysis, with the latter being more costly and time consuming but having the advantage that scientists gain an inventory of precious samples.

Only after these steps are completed will it be possible to make a well-informed and justified decision about whether the object is economically viable for space mining. This is where space-mining companies will be breaking new ground, so to speak, since space agency missions have barely gone beyond the phase of initial exploration and characterisation. The result is that any step past this phase has not been tried in space before and will, therefore, require significant investment in new technologies or, at the very least, extensive development of current technologies to make them fit for purpose. The limited global experience we have in the advanced space-mining steps means that the return of mined materials to Earth, whether processed or not, or their use in space, is still a long way off, probably decades at best.

Who is already in the game?

What we should remember is that space agency missions that have come and gone, are on their journeys or are set to begin work in the coming decades, have collectively consumed billions of dollars of investment. These missions were in development for exceptionally long timeframes in order to achieve demonstrably suitable and reliable technologies. To make space mining economical, the companies involved will need to build quickly on the lessons learnt from these space agency missions and, in some cases, pioneer new and perhaps more efficient ways of working.

Some of the companies that have declared an interest in space mining – to much fanfare in the media – are already making progress towards their lofty goals and are focusing on the initial steps discussed here. Satellites are being built that can be used for prospecting, detecting water and minerals in space. Alongside this, companies are looking at ways to reduce the cost of launches and, importantly, trying to find a way to generate revenue while they wait for the space-mining industry to take the leap of faith required.

The company Deep Space Industries (DSI), originally set up in California, USA, began working in collaboration with the government of Luxembourg on a spacecraft called *Prospector-X* to operate in low Earth orbit. *Prospector-X* was designed to test navigation and propulsion systems and avionics technologies, to pave the way for the next generation of spacecraft, which might actually be the ones to search for potentially lucrative asteroids. DSI is also working on a number of small 'laptop-sized' spacecraft that will progressively be sent to explore near-Earth asteroids on one-way missions. It is said that such apparently cost-effective spacecraft will be able to perform reconnaissance, detailed analysis and mining activities on separate missions.

Planetary Resources, another company with a focus on asteroid mining, was set up in Washington state, USA, backed in part by high-profile names such as filmmaker James Cameron and Google co-founder Larry Page. Planetary Resources has been designing several satellites to be used in low Earth orbit as part of their Arkyd series. Each step in the Arkyd programme is designed to incorporate progressively more developed systems capable of more complex tasks. Earlier models will act as telescopes to scout out suitable asteroids, specifically the *Arkyd-100*. Subsequent models, such as the *Arkyd-200*, are the 'asteroid interceptors', carrying a suite of instruments to an asteroid that is passing close to Earth, in order to analyse its physical properties. The *Arkyd-300* would contain a more advanced propulsion system to travel into deep space to intercept those asteroids that won't be passing close to Earth, and also analysing their physical properties and chemical composition. Being able to approach and intercept suitable asteroids in near and deep space to collect scientific data on their physical characteristics and to conduct sample analysis is a key step in the mining process as outlined previously. Planetary Resources has already taken the first step by launching the *Arkyd 3 Reflight* into low Earth orbit from the International Space Station in 2015. This was a small, successful demonstration satellite to test avionic and propulsion systems that will hopefully be used on larger satellites later in the Arkyd series.

Even though the primary goal for Planetary Resources is to mine asteroids, it will take some time to get through the initial steps of any space-mining programme. In the meantime, the company aims to make the most of its developmental steps by making its space telescopes available for private hire. Adding to this, it plans to further develop a system of satellites called Ceres, based on the Arkyd spacecraft, that can be used to look back at Earth from space, instead of away from Earth at asteroids in deep space. Such a system could be used initially for Earth observation to provide hyperspectral imaging of the planet – effectively going beyond what the human eye can see and representing a step up from satellite imagery – which is said to provide on-demand information on our planet's natural resources. This technology could be used in agriculture, environmental monitoring and by energy industries. Although this may seem like a side-step away from asteroid mining, these techniques and spacecraft will eventually be used in the exploration and prospecting phases of space mining. In the meantime, such endeavours will be a useful money-spinner, allowing Planetary Resources to stay in business as it works its way towards the ultimate goal of successful and profitable space mining.

Tug and park – driving space objects

Contrary to what the name suggests, NEOs are not always that 'near' to us because, although their orbits do indeed bring them close to the Sun and Earth, the result is that they also spend a large proportion of their orbit very far from the Sun. As we've seen above, there are many complications with the operation of complex scientific instruments aboard spacecraft far from Earth – long communication times and lack of sunlight to harness power being just two. As we know, the *Rosetta* spacecraft had to enter hibernation for many months to save power on its way to catching comet 67P/C-G. The communication-delay time during this mission was up to 30 minutes each way, depending on the distance from Earth, which was fine in the circumstances because the *Rosetta* teams didn't need to communicate with their sleeping beauty at this stage. A long delay in communications would have made things more complicated if the spacecraft had been expected to perform analyses or mining activities at the time.

Of the space-mining activities that follow the prospecting phase, those involving resource extraction and processing of extracted materials are much easier if the object is closer to Earth. Therefore, chasing, capturing, tugging and parking an NEO in a location that is more easily accessible for subsequent spacecraft visits seems like a sensible plan. It turns out that NASA thinks this idea is achievable, and has already made plans to start practising such a manoeuvre with its Asteroid Redirect Mission (ARM). Initially NASA doesn't plan to capture an entire asteroid, but has suggested kidnapping a small piece of one, a multi-tonne boulder probably 2–3m (6.5–10ft) in diameter, using a robotic arm attached to a spacecraft. The asteroid NASA has selected to visit for this purpose is 2008 EV5, a 400m (1,312ft) diameter asteroid, although asteroid Bennu, the target for the *OSIRIS-REx* mission, and Itokawa, which was the target of the *Hayabusa* mission, are also potential options. NASA plans to tow the kidnapped boulder to a so-called 'distant retrograde orbit' around the Moon, to 'park' it until a follow-up mission can be launched to investigate it in more detail.

If NASA is successful in its initial endeavours, then it wants to capture larger objects that could be investigated in the future by manned missions. This means that ARM could act as a first step towards putting humans back on the Moon, or even on journeys to Mars. Initially, however, the plans call for robotic instruments to start the job and, once the scientific studies are complete, the idea is that private space-mining companies would be invited to access the boulder to test out their own technologies. If NASA and private companies can work collaboratively on such missions, then it is likely that they can achieve more, and in a faster timeframe, by learning from each other. However, such plans have been criticised by some who suggest that ARM hardly represents groundbreaking science when we've already returned materials to Earth from outer space, and that it would be better to invest in technologies that can work on asteroids in their natural orbit. After all, if we can build a spacecraft to get humans to Mars, similar technologies would eventually be capable of exploring asteroids.

Additionally, the idea of towing rocks around space is, at the time of writing, totally untested and could pose some serious risks. One of these is that NEOs vary widely in composition and structure, so while towing and repositioning a highly consolidated chunk of asteroid may turn out to be very simple, doing the same with a poorly consolidated crumbly, icy piece of primitive asteroid or comet might present some problems. For example, what if the object crumbled into pieces in transit, or once it was placed in a new orbit? Furthermore, one of the reasons NASA's initial plans don't involve moving a large asteroid to a new orbit is that there is always the risk of a life-threatening disaster if the asteroid was inadvertently given a trajectory towards Earth. Should things go wrong for ARM initially, a small, boulder-sized object would be expected to burn up on atmospheric entry to Earth and, therefore, would pose no risk to life. This approach represents a relatively safe way to test new technologies. However, the real test comes if, and when, such plans are scaled up to larger objects. The obvious future risks associated with such endeavours will need very careful consideration by government agencies worldwide. Only time will tell whether ARM will be possible and, if so, whether it will be a costly, but ultimately wasted, endeavour.

Mining asteroids for metal

As we've seen, a small M-type asteroid totalling just a couple of kilometres in diameter has the potential to be worth trillions of US dollars, depending on the exact mix of metals it contains. In most metal asteroids, it is iron and nickel that are the most abundant, composing up to 99.9 per cent. This is known from the analysis of iron meteorites on Earth as well as remote study of the asteroids from which they potentially originated. However, when it comes to the rarer metals such as platinum, osmium and palladium that compose less than 1 per cent of these objects, even though they may occur in much lower abundances, their rarity on Earth and consequent high market value means they could be worth the same as all the iron and nickel put together.

The so-called 'rare earth elements', which are all metals and thus are sometimes referred to as the 'rare earth metals', are a key focus for asteroid mining. They are metals that you might not have heard of before, except in your school chemistry lessons, such as cerium, dysprosium, neodymium and yttrium, to name just a few. These elements are not actually as rare on Earth as their name suggests, but instead got their name for being a little bit difficult to extract from their host compounds, which are typically oxides ('earth' in this context is now an obsolete term for oxide). Such metals commonly occur within the Earth's crust but tend to be dispersed and mixed with each other, often present in low abundances that aren't economical to extract. However, their use on Earth is widespread, with one of the main examples being that they are used in rechargeable batteries that are extensively exploited by many electronic devices such as phones, cameras and portable computers, but also in increasingly important electric vehicles. These rare earth metals are in obviously high and increasing demand and our reliance on them means that a long-term supply is of paramount importance for modern society as we know it. It will still be possible for the Earth's resources to meet the demand for these elements for the foreseeable future, at least decades, but their extraction will become increasingly expensive over time as the more easily accessible and abundant resources are depleted. The remaining resources, the more dispersed ones, will require much more intensive effort to extract, which will not only be more expensive but could also result in hugely devastating environmental effects.

It is a similar situation for the platinum group elements (PGEs), another group that is of interest for asteroid mining and hard to obtain on Earth. The PGEs are a set of six metals that sit in the middle of the periodic table: ruthenium, rhodium, palladium, osmium, iridium and platinum. PGEs are siderophile elements, meaning they are 'iron-loving', as opposed to lithophile elements, which are 'rock-loving', hence they combine easily with oxygen. The PGEs' affinity for iron means that, when the Earth formed, they tended to preferentially gravitate towards the Earth's core because it is itself made mostly of iron. The result is that these elements are not very abundant in the outer layers of the Earth – the bits that are easier to access for mining – and despite what Hollywood may suggest, we can't drill down to the core to obtain them. The PGEs have many uses on Earth because they are highly resistant to wear and chemical attack, have excellent high-temperature characteristics and have stable electrical properties. This means that there is a high demand for them each year for use in a wide range of products including, but not limited to, electronics, medical equipment, glass, turbine blades, catalytic converters and jewellery. Low supply and high demand – being found in one in four goods that we use on Earth every day – means that they come with a high price tag.

Despite platinum being only a minor constituent of asteroids, making up just 20 parts per million, which sounds like a rather minuscule amount, this still means that a 500m-wide (0.31 miles) asteroid could contain more platinum than has ever been mined on Earth. Such an object could potentially have enough platinum to last the next 200 years at the levels of current demand. However, if a large cargo of a precious metal were suddenly returned to Earth, a metal that is only expensive because it is rare, the risk is that it would cause a massive price drop, making the whole process of obtaining it instantly uneconomical.

PGEs are found in a few restricted locations of the Earth's crust, with one example being the Bushveld Igneous Complex in South Africa. This source will be available economically for a few more decades and there may be more, as yet unproven, deposits in the region that could extend the supply for up to a century. However, this doesn't present a long-term solution, and requires vast, and often ugly, mining operations in areas of natural beauty. Hence, obtaining a source of these elements from asteroids might be a more viable long-term solution, meaning that we could maintain Earth's natural beauty yet still obtain the resources we need to sustain our development of advanced technologies.

The problem is that extracting these precious metals from an asteroid is not a straightforward process, partly because they are alloyed to the iron and nickel present in the space rock. It's possible that a chemical separation technique such as the Mond process could be used: this would vaporise away the iron and nickel to leave behind a residue of the remaining siderophile (iron-loving) elements, which would include the platinum group and other rare metals. However, to further separate these from each other would probably involve sending the ore back to Earth. This would be good for our terrestrial technology, but we might also want to use these metals in space, to build tools and space equipment and develop space infrastructure. This means they would need to be launched back into space after secondary processing on Earth, a potentially costly business. Initially expensive processing techniques will be worthwhile, however, due to the value of the metals in question. With restricted supplies of precious metals on Earth, technology developments that rely on the use of these metals are going to become somewhat limited and expensive. Therefore, the prospect of an almost unlimited source of these elements in space is such an attractive one that it could lead to the development of inexpensive and new technologies.

Mining water – a universal solvent as useful in space as on Earth

Of course, metals are not the only precious materials that comets and asteroids contain. Some space objects are also expected to contain high abundances of non-metals such as oxygen, hydrogen, carbon and nitrogen that all have societal value. If commercial space-mining companies are going to the trouble of capturing an asteroid or comet, towing it somewhere and mining it, then it makes sense to process the entire thing, to separate all the constituent ingredients from each other.

Despite its abundance on Earth, water might be one of the most important, and useful, resources that could be mined in space. It definitely won't be overlooked in favour of the seemingly more lucrative precious metals that space also has on offer. Water is essentially the fuel of space. After all, rocket fuel is liquid oxygen and liquid hydrogen. Of course, these elements would need to be separated from each other if water mined from space objects were to be used. Recently, however, technologies have been developed that could simply use superheated water as a propellant, to avoid having to 'crack' the water into its constituent parts, something which is actively being investigated to make 'green' energy.

Either way, the use of mined water as fuel in space should be the next priority and is a key step in the space-mining process. Without a local fuel source, space-mining efforts will struggle to be economical because the techniques employed to collect and process resources from space objects are expected to be highly energy intensive. If a commercial organisation focused on mining and storing water in space, they could make it available in a similar way to a roadside 'filling station' on Earth. Such a propellant depot could also be used as a satellite-servicing area, extending the lifetime of satellites used by governments or private companies for defence or communications, when their orbital manoeuvring fuel is low. Solar energy is, of course, in plentiful supply in the inner Solar System, but it requires vast swathes of solar panels to meet the demand of a spacecraft filled with instruments, and solar energy cannot be relied upon during missions into deep space.

The set-up costs for the commercial mining of water in space will be large but, luckily, it turns out that water isn't just useful as fuel in space, it has many other applications. Thinking longer-term, mining water in space would have obvious applications for any space-based human activities. Humans and any other animals or plants simply can't survive without water. Water is required for drinking, hygiene, irrigation and it could even be used to produce oxygen for breathing, as well as a radiation shield. Minimising crew exposure to radiation in space during long-duration missions is a major challenge that must be overcome before we send humans off to explore deep space. Radiation in space comes mostly from the Sun, which fires out high energy-charged particles in the form of solar wind in all directions into space. There is also a second source of radiation that comes from galactic cosmic rays – particles that are accelerated to near the speed of light from other stars in the Milky Way or galaxies. Both these types of radiation are predominantly made of protons and can have highly damaging effects on cells, irreversibly breaking apart DNA and causing mutations as they pass through, which could eventually lead to death after prolonged exposure.

On Earth, we are fortunate to escape the hugely damaging effects of these particles thanks to Earth's immense magnetic field, which acts to divert the charged particles away from us and back into space, or block them completely. In fact, without the magnetic field we wouldn't have our atmosphere either, so we have another reason to be thankful it's there. Even the atmosphere itself is useful as it absorbs the vast majority of particles that do happen to make it through Earth's magnetic protection bubble. Conversely, the Moon, Mars, comets and asteroids either lack both these protection shields or they don't work as well.

Before even getting to an extraterrestrial destination, simply travelling through open space is a big problem because of a lack of protection from radiation. The astronauts aboard the International Space Station (ISS), despite living outside of our atmosphere, are still afforded some protection from radiation because they are within the Earth's magnetic field. However, studies suggest that they receive about 1 millisievert of radiation a day, which is equivalent to a year's dosage of radiation on Earth, with a sievert being a measure of the biological effect of ionising radiation. A human receiving a dose of just 1 sievert will experience a 5.5 per cent increase in their risk of contracting cancer. The plus side for the ISS astronauts is that they don't tend to stay on the space station for very long, and since biological radiation damage is calculated as a cumulative risk, these doses are deemed acceptable. However, a six-month or even one-year stay on the ISS is not nearly as bad in terms of radiation dosage as a multi-year excursion to Mars. The *Curiosity* rover that is currently on the surface of Mars measured a radiation dose of 0.66 sieverts during its 253-day cruise to the red planet. This is a rather worrying amount, so high in fact that astronauts travelling to Mars could die on their way there. Clearly, we need to find a way to protect our future space explorers, otherwise we're not going to be able to safely explore very far from Earth.

Luckily, apart from magnetic fields and thick atmospheres, there are other ways to block or divert harmful radiation. The ISS has a polyethylene shield that cuts out a few per cent of an astronaut's radiation exposure. However, it is calculated that the thickness of polyethylene required for the protection of a spacecraft transiting into deep space, or parked in a Lagrangian point where space mining could take place, would be prohibitively heavy since it would be the *only* protection from radiation. The cost of transporting these shielding materials from Earth would also be high.

If humans were living on a lunar or Martian base, then their living quarters could be protected by a shield made of regolith – a layer of rock covering the buildings. This regolith wouldn't require transport from Earth as local rock could be used, but equipment would be required to dig it up. A regolith shield is obviously, however, not a practical solution for a spacecraft travelling into deep space. There are other potential solutions, including the slightly whacky option of bagging up astronaut excrement to fill the walls of a spacecraft, but they aren't very appealing. Luckily there is one element that is excellent at blocking protons, and it is hydrogen. Not only is hydrogen one of the most abundant elements in the Universe, it is a substantial component in many common compounds, most obviously and notably water. For water to afford humans embarking on long-duration space journeys protection from radiation, there are plans for spacecraft to be designed with an outermost water-filled layer to protect the precious cargo within. Launching spacecraft from Earth pre-filled with water for this task would be technically challenging – it's too heavy – so being able to obtain the water in space post-launch could make these long journeys feasible. A pit-stop at a handy water-rich asteroid or comet to obtain this water, or at a propellant depot, might be an easier solution. It is early days and there may be other solutions to the radiation problem but, whatever the case, mining water in space will be of fundamental importance for any future missions.

While some NEOs are expected to contain a high level of water, up to or even exceeding 20 per cent, one of the problems is that this water is not always very accessible. The water in some NEOs is essentially 'locked up' by strong chemical bonding within rock minerals. Such bonds require significant amounts of energy to break, making it potentially costly to free the water from its host rock mineral jail. This scenario is more common in the asteroids, where the water has often been 'processed' into the rock during the relatively complex earlier life of the asteroid. This means that even though these objects can potentially contain a significant amount of water, getting it out of the rock is not straightforward.

Fortunately, not all water in NEOs is hard to access since comets, and some asteroids, can also contain a lot of free water in the form of ice that would be much simpler to harvest. Even the Moon has water ice present at its poles, but, as we've seen, it would be costly to lift this water off the surface after mining. Thus, comets and asteroids are a more attractive prospect gravitationally. Additionally, the water ice present in space might be joined by ices of other volatiles, such as methane and carbon dioxide, which could also be useful for space-mining activities and fuel generation.

One suggested method to release water from small asteroids or comets involves literally bagging up the captured object and harnessing highly concentrated sunlight to 'sweat' the water out, by using the Sun's energy to break it up and/or 'dig' holes in it to release water and other volatiles. This 'sweating' technique is known as 'optical mining', concentrating solar thermal energy using concave mirrors, and is something that is initially being tested on Earth prior to deployment in space. If successful, optical mining has the potential to deliver hundreds of tonnes of water from one small asteroid into a near-Earth orbit using a fully autonomous spacecraft, since it's not a particularly complex task. One of the main challenges will be finding a way to separate ices that are mixed with each other within a space object, and then store the separated volatiles once they are mined. As with other issues, maximising extraction in this way will require investigation once space mining becomes a serious prospect.

Human role in space mining

A key area of consideration in space mining is whether humans will be required, or whether all the activities can be performed remotely using robotic spacecraft. As we've discussed, initial mining efforts are more than likely to be carried out robotically. Allowing robots to take the burden of the risk, and the hard work, is a practical starting point for any initial forays in space. Chasing down far-flung, speeding asteroids is a major challenge on its own and would be significantly more expensive with humans, not to mention more technologically and biologically risky than using more easily expendable robotic spacecraft. Sending humans to explore our Solar System continues to be a dream, one that humankind has been working on for many decades but has achieved on only a handful of occasions – ignoring the many trips to the International Space Station, of course. More recently, some exciting proposals have been discussed for sending humans to Mars and initiating a human colony on the red planet. Mars One is one organisation interested in this cause. However, even it admits there is still much to achieve before its plans can become reality, despite the apparent hordes of people who have volunteered to take the possibly one-way journey to Mars. These humans will require robots to have paved the way for them, mining and preparing the resources they will require to live, and sustain their lifestyle, in space.

It may not be instantly obvious how mined metals could be used by humans in space. After all, on Earth we tend to think of humans controlling large factories to make all the objects we use in everyday life. However, this is where the technology for 3D printing could play an important role, with the first space 3D printer up and running on the International Space Station in 2014 currently testing the technology. The idea is that astronauts on long-duration missions would be able to simply print anything they required, so long as they had the right printing supplies for the printer: ingredients such as metals that could come from asteroids or comets. Without an advanced 3D printer, anything astronauts require in space has to be taken with them, or sent up later using a resupply mission. As we've discussed many times, launching materials into space is expensive but, more importantly, it can take weeks or even months to launch a resupply mission; the further away the astronauts are from Earth, the longer it will take to get the supplies to them. I think the book and movie of *The Martian* demonstrated this problem really well. The waiting time for a resupply mission might be too long if the astronauts were to find themselves in a dire situation with broken equipment which they didn't have the required tools to mend. Even something as simple as a broken toilet could easily spell disaster for astronauts living in an enclosed habitat. Space exploration is a bit like preparing for a camping trip – you simply can't predict everything you need for the journey, despite your best efforts. When camping, you have the option of giving up and going home if, for example, your camping stove breaks. This simply isn't possible for space missions visiting far-flung parts of the Solar System, where broken equipment could result in a life-or-death situation. 3D printing any desired tool or spare part could literally be a lifesaver.

If we, as humans, really have the ambition, or even the requirement one day, to explore beyond our exquisite world, then we must establish the capability to live and survive in space, for lengthy periods or even lifetimes. This will almost certainly rely on obtaining resources from other space objects we encounter along the way, as we can't rely on bringing everything we need from Earth for the exploration of deep space, much as our ancestors tried when exploring our unknown world centuries ago. In this respect, space mining is the first step in creating human civilisations in space, however far off this goal might be. We must learn to utilise the resources that are out there, which are sometimes in plentiful supply, instead of forever relying on our own planet to fuel and equip our journeys into the unknown. The result is that we must invest in human space exploration at an early stage, and new endeavours in space mining, in order to guarantee our readiness when the opportunity presents itself for humans to meet the goal of exploring beyond our planet, or living on another.

Who owns space rocks?

An interesting problem facing anyone attracted to the idea of mining space for profit is the issue of who owns space and, therefore, who owns the mined products. What happens if a private company wants to exploit our nearest asteroid to return rare metals to Earth for sale? Is it a case of finder's keepers? Well, if you're an American citizen there's nothing to stop you from going into space, catching an asteroid and hauling back some rocks to Earth. Well, OK, there's quite a bit to stop you, but you are allowed to, theoretically, own the rock you collect. In 2015, a key milestone was made when President Obama signed the US Commercial Space Launch Competitiveness Act which recognises the rights of a US citizen to own any space resource they obtain from an object. Although the law states that no nation or person can claim ownership of another 'heavenly body' thanks to the UN Outer Space Treaty of 1967 – the Magna Carta of space law – it doesn't prevent US companies from exploiting them and extracting any materials they find.

The USA was the first country to pass a law on such activities and this move was met with some criticism. It was seen as a flagrant misinterpretation of the original UN Outer Space Treaty, which was set up as a sort of peace agreement to prevent the placement of weapons of mass destruction in space, and forbids any government from staking a claim over a celestial body. However, in 2016 the government of Luxembourg announced that it would create a legal framework to 'jump-start' an industrial sector to mine asteroid resources in space and it is likely to follow up with a similar law to that passed by President Obama. Luxembourg also announced that it would provide incentives for space-mining companies and was the first country to provide direct capital investment to a space-mining company, partnering with Planetary Resources to develop technologies and services required for space mining.

It seems likely that many other countries will follow the USA and Luxembourg in creating their own laws, frameworks and incentives for private companies interested in the exploitation of space resources as, although the economic viability of space mining is still unproven, the risk of not being involved in such an endeavour is too great. However, with the new developments in space exploration and the potential for mining in space, the UN Outer Space Treaty is increasingly inadequate. Whether it becomes redundant over time is hard to predict, but in the fast-moving and groundbreaking field of space mining, new laws will need to be drawn up quickly. International law is, understandably, complicated, particularly with respect to space mining. It seems fair that everyone should have an equal and fair bid for resources in space, under conditions that are agreeable to all on an international scale – not something that is going to be easy in a competitive and politically charged world. Perhaps we only need to consider the present-day race between powerful nations to mine Earth's deep oceans and poles to see the potential for problems ahead.

The exploitation of space has been likened to fishing in international waters. The fishermen don't own the fish until they are on the decks of their boat, and they don't own the waters they fish in, but they have the right to cast their nets in those waters. This makes fishing laws seem simpler than they are, and they are logistically hard to police, so it can only be expected that enforcing laws in deep space is going to be even more complex. Ultimately, international cooperation on a very grand scale is going to be fundamental to the success of space-mining endeavours. This is required because there are some considerable risks, as well as potentially large rewards for future generations. Surely, it would be immoral for a few private companies or governments to have a monopoly on our Solar System's resources, but since this is the precedent that has been established over much of the Earth's resources, it is hard to know what the future holds in space.

Another issue is whether we are morally comfortable, as a human race, with destroying parts of space for commercial gain. Many people think that we've ruined Earth enough already, so maybe we shouldn't extend this to the rest of our Solar System. Alternatively, it might get to the point where we could protect Earth from further damage by mining objects in space. One of the advantages is that we probably wouldn't even notice these objects disappearing as they're so small, with many hard to find even with the best telescopes from Earth, and in such plentiful supply. Such an approach does, however, have a feel of NIMBY (not in my back yard) about it – protecting the Earth but destroying the space around it.

Of course, the effects of space mining on the Moon might be more obvious. Excavations on the surface of our closest neighbour could leave it visibly scarred to even an amateur with a telescope, but this would require very intensive and large-scale mining efforts, which seem unlikely given the complications. Companies extracting resources on Earth are expected to comply with legislation that protects the environment and heritage of the area they work within. Whether this is achieved in many cases is probably open to argument, but there is currently no such agreement to cover space-mining operations. Ultimately, for most people, the debate is poorly understood and will, at least initially, be controlled by the economics of the situation. The higher the demand for these metals, and the less the supply from Earth as resources diminish, the more inevitable it is that some asteroids, and possibly the Moon, will be exploited. Only time will tell what the reaction to these activities will be among the general population on Earth.

It is hard to say when space mining will become a reality. There are a lot of technological and logistical wrinkles to iron out before we see mining in space become something we can take for granted. With the likes of private space companies such as SpaceX and Blue Origin slashing the cost of launching bulk into space, we are entering an exciting new phase of pioneering space exploration. However, the cost of space launches will still need to come down drastically if we are to see economical transport of resources from space back to Earth, if indeed that is something we want. As it is, current space-mining activities will no doubt pave the way for humans to explore further beyond our beautiful planet and lunar neighbour, and surely this is one of the most exciting prospects about the entire complicated endeavour.

CHAPTER 10

Mission 'Save Planet Earth'

The United Nations has designated 30 June as International Asteroid Day, which to many people may seem like a strange thing to do. It certainly isn't because asteroids are about to become extinct, like some endangered wildlife. Instead, it's because there's a threat that we, as humans, could become extinct if an asteroid were to collide with Earth. The prospect of our planet experiencing a devastating, life-destroying impact by a comet or asteroid may sound highly unlikely, but it is something that is almost certainly going to happen at some point in the future. The question is when?

For those of us already here on Earth, luck seems to be on our side for now, with scientists predicting that we should be safe for at least the next 100 years as there are no NEOs predicted to be on a direct collision course with Earth in this time. Nevertheless, there is always the possibility a random object that scientists can't yet see is lurking out there in the outer Solar System, in an orbit that intersects that of Earth within the next few decades. The problem is that it is currently impossible for astronomers to track every object in the Solar System, particularly the small and fast-moving ones that are on random orbits. Despite this, you may wonder why we should care about the Earth experiencing a large impact, particularly one that might happen more than 100 years from now, at a time when very few people alive today will still be around. However, it's not a comforting thought that we may leave behind for our descendants a planet that's ill-prepared to deal with a potentially cataclysmic, species-killing impact from space, particularly if we can work together now to do something to prevent it happening. Hence, Asteroid Day – the date of which marks the anniversary of the Siberian Tunguska event – is designed to get us thinking carefully about what our global plans will be if we discover that an object is heading straight for us.

An impact from space could have globally devastating effects, so working out how to deal with it involves enormous international cooperation. We just need to look at the huge scar that comet Shoemaker-Levy 9 left in the side of Jupiter, when it careered through its atmosphere in 1994, to wonder what the same impact could have done to Earth, which has a volume 1,300 times smaller than Jupiter. Luckily, scientists and governments have already begun working on ways to protect our planet from such space threats, but it is still early days. This area of interest comes broadly under the term of 'planetary defence' and includes the science behind observing and tracking NEOs, as well as focusing on how to either divert or destroy an object should scientists identify one on course for our planet.

The impact of an impact

If we fail in the future to protect the planet from space threats, then we could expect a large asteroid or comet impact to wreak havoc on the Earth's surface, resulting in major global changes and high death tolls. It wouldn't just be humans that would be affected, of course, but many other Earth species, too. A comet or asteroid impacting one of Earth's oceans would result in huge tsunamis that could utterly devastate surrounding coastal regions. It's anyone's guess how many millions of people would be directly or indirectly affected by a tsunami radiating out from an impact location in the middle of the Atlantic or Pacific Ocean. If we just imagine the ripples radiating from a pebble being thrown into a pond, with nothing to get in their way, the ripples would travel as far as the pond's edge, and although they wouldn't be expected to demolish the pond's borders, depending on the size of the pebble and the force with which it entered the pond, we could expect water to flood out over the edges once the ripples reached that far. Now, let's imagine scaling up to a huge chunk of asteroid, possibly a few kilometres in diameter, travelling at unimaginable speeds – up to 65,000kph (40,000mph) – into one of our oceans. The effect is the same, but the edges of the pond are now large cities inhabited by millions of people, and dense infrastructure that certainly couldn't cope with being inundated by a massive wall of water. A large comet or asteroid strike on land wouldn't be any better, swapping water ripples for ground waves, and resulting in huge shockwaves that would easily flatten buildings, at the same time as ripping up and melting the ground around the impact location.

As we previously saw, the dinosaur-killing impact was probably produced by an object just 15km (9 miles) in diameter, and while it didn't manage to destroy the entire planet, it certainly wasn't good news for the dinosaurs or half the species that were around at the time. In fact, a similar impact at the present day would eventually kill off all humans, if not at first then within a few years. Such a collision could easily instigate long-lasting, global catastrophic effects on the planet's biosphere, brought about by the initiation of an impact winter, where the sun's rays are blocked out by the encircling dust and debris kicked up from the collision. Under these conditions, any life that managed to escape the direct effects of the impact would struggle to find enough food and resources to survive for long after.

It's not just the massive asteroids and comets we need to worry about, because even modest-sized objects, on space-scales – maybe those no bigger than a suburban detached house – could have exceptionally devastating effects locally. Space objects can be travelling extremely quickly, so even something house-sized would flatten everything in a 1km (0.6 mile) radius from the impact location, including sturdy concrete buildings.

Moving up to slightly larger objects, there is an NEA, 1997 XF11, which is about 1.5km (1 mile) in diameter and is set to make a close pass to Earth in 2028. In fact, initial predictions suggested it might collide with Earth, but after some careful observations scientists have, fortunately, worked out it will pass at about 2.5 times further than the distance from the

Earth to the Moon, around 930,000km (580,000 miles). Despite this seemingly huge distance, it isn't actually very far in the great cosmic billiard table of the inner Solar System so, although it won't cause us any concern, with a zero probability of collision with Earth, it is closer than many large objects have been recently and is being used as a thinking exercise in case something similar is on course for Earth. It is important we track objects in this size range, because if they were to collide with Earth then they would be expected to flatten everything within an 800km (500 mile) radius. Put another way, if this asteroid were to strike London, UK, then the area of devastation would incorporate the major UK cities of Manchester, Liverpool and Cardiff. It would also affect parts of France, including Paris, and other nearby countries such as the Netherlands and Belgium.

We've discussed comets a great deal here and so you might be wondering how an object, however large, made almost entirely of ice and fine-grained, fragile dust is going to cause much damage if it impacts Earth. Surely the effects of atmospheric entry will be more than enough to destroy it completely? Well, you'd be partly correct in thinking this, as a 'classic' comet structure probably wouldn't survive atmospheric entry to make it to the ground, at least not in one piece. However, if a large incoming comet was destroyed in the atmosphere, then it could be expected to cause an extremely large explosion, so powerful that it could have the same effects on the Earth's surface as would a massive earthquake. If we think back to Chapter 3, where we introduced the powerful Siberian Tunguska event, this was an air blast caused by an incoming meteor that was powerful enough to flatten dense forests across a 2,000km^2 (770sq miles) swathe, yet no fragments of space rock were found on the ground. Although there are various estimates, the object that caused this was thought to have been only 50–200m (160–650ft) in diameter, teeny on Solar System scales, and may have been a comet, or something of similar structure. Then there is also the modern example of the Chelyabinsk meteorite, a relatively small meteor that exploded on atmospheric entry in an impressive air blast, scattering meteorite debris across a wide region. Except it wasn't the rock that posed the greatest, or any, risk to humans. Instead, the air blast itself was the reason for people reporting injuries, as they were hit by flying glass and debris created by the explosion.

Although scientists are keeping a good eye out now, and tracking many hundreds of NEOs, it is a statistical certainty that a life-destroying space object will be on a collision course with Earth in the future. We've learnt that there is an inverse relationship between the size and frequency of impacts. On average, rocks as large as 4m (13ft) impact the Earth once every year and do little damage, mostly disintegrating on entry with no large air blast. Larger objects hit much less frequently but when they do, the effects are more devastating. It is these larger objects, those above about 100–150m (330–490ft), that require further action if we should identify one heading our way, so that we prevent a humanitarian disaster on Earth. This implies that we must divert or destroy the offending object before it gets to us. It's all well and good knowing if we are in the path of an object and that we want to move or destroy it, but we need to figure out how this can be achieved. Even if an impact was about to occur on the opposite side of the planet, the chances are that the entire globe would be affected in the ensuing days, weeks and months, hence the need to make this issue a global concern.

Working out where everything is and how likely it is to collide with Earth

The way we will deal with an object heading our way will very much depend on how far away it is when we figure out it's on a collision course with Earth, what size it is and what we think it's made of. As we saw in Chapter 2, scientists are getting quite good at detecting NEOs, particularly with the Spaceguard Survey. By 2011 they estimated that they'd found around 93 per cent of the NEAs that are 1km (0.6 miles) in diameter or greater, and are aiming to have detected

90 per cent of the NEOs down to those that are 140m (460ft) in diameter or greater by 2020. There is even a project called ATLAS (Asteroid Terrestrial-impact Last Alert System) that is designed as a 'late-warning' system, searching for those sneaky objects that could be due to impact Earth within one to three weeks. Of course, it's unlikely we could do much about these objects in time to divert or destroy them, but it would allow us to instead work on evacuating and preparing the forecasted target region in the same way we would for a predicted volcanic eruption.

For the less sneaky of objects, those that announce their upcoming arrival to the near-Earth region well in advance, it is projects such as NEOWISE that are capable of finding out more detailed information about them, such as their diameter and albedo. These are characteristics that are crucial for scientists to understand as, although the broad motion of asteroids and comets is fairly straightforward to predict, being controlled by gravity, it turns out there are some subtle non-gravitational processes that they must get a grasp of in order to predict the exact trajectory of an object to within metres rather than kilometres. It is particularly important that scientists assess the effects of these non-gravitational processes on objects that are predicted to pass very close to Earth, the so-called Potentially Hazardous Objects (PHOs), where a small change in their position could be the difference between a near-miss and an impact with our precious planet. Incidentally, the PHOs are defined as having an orbit insertion distance with respect to Earth of less than 0.05AU, or around 19.5 lunar distances, and to have a magnitude of 22, a measure of their brightness which corresponds to their size. This measure is defined as the size at which an object could cause major global devastation, around 100–150m (330–490ft). As we don't know the exact shape, albedo (light reflectivity of a surface) and composition of some of these PHOs, they must remain on the list until scientists study them in space and better refine their characteristics to calculate how the subtle non-gravitational phenomena will affect them, and whether their orbits will intersect with our planet.

Two of these non-gravitational phenomena are known as the Yarkovsky and YORP effects. The Yarkovsky effect, named after the Russian civil engineer Ivan Osipovich Yarkovsky, who discovered it, considers that variations in albedo across the surface of a small object in space can act to increase its rotation rate. Basically, as an object is warmed by the Sun, it re-radiates heat, which can create a tiny amount of thrust. Because of the relatively small force that is produced by the Yarkovsky effect, it is more noticeable on small objects – meteroids and small asteroids around 10cm to 40km across, the upper end of which you might not consider particularly small. Even a tiny momentum acting over millions or billions of years can produce a significant change to the orbit of objects in this size range.

YORP, an acronym combining four scientists' names – Yarkovsky, O'Keefe, Radzievskii and Paddack – is a second-order effect of the Yarkovsky effect and relates to how the shape of the object can influence its rotation, having a sort of windmill effect that can act to change the object's rate of rotation. It's easy to imagine that a perfectly spherical object might rotate differently to something with a more complicated geometry. Just think how a football and a rugby ball differ in their motions through the air. When we think about comet 67P/C-G – an object formed of two lobes and affectionately named the rubber ducky – it illustrates very well the potentially weird shapes of small space objects. In fact, the smaller space objects are rarely spherical, since a sphere only tends to form as objects become planet-sized and can gravitationally pull themselves into that shape.

Apart from affecting the orbit of a small space object, another complication of the Yarkovsky and YORP phenomena is that they can act to break apart an object, literally causing it to spin itself to pieces. In fact, in 2013 an asteroid named P/2014 R3 was observed to have done exactly this due to its increased spin rate brought about by the YORP effect as it neared the Sun. In some ways, it might be useful if a hazardous Earth-crossing object were to destroy itself before it got to us, but it would be hard for scientists to predict this, and even if they could it would result in an 'edge of your seat' ride, waiting for it to break up before it reached us. Scientists are currently working on refining the Yarkovsky and YORP effects but need to understand the shape and thermal properties of the object in question, otherwise it's impossible for them to carry out the calculations required to predict its orbit. As such, there are many small objects that will remain on a potentially hazardous orbit list until more can be learnt about them.

So, we've decided we're in the firing line … what next?

If scientists could predict for certain that a space object was on a collision course with Earth then we would have, hypothetically, a few options available to us. The mitigation techniques can be placed into two categories that are either precise and gradual – to divert the object – or less precise and sudden – to destroy the object. The use of any technique to prevent an object colliding with Earth gives humans the opportunity to shape the Solar System to enhance our chances of survival, but we must act carefully to avoid making the situation worse. The technique chosen would depend on what was known about the object heading our way, its composition and structure, and how long we had before it was due to impact Earth.

The best time to divert or destroy an object would be when it was still very far from Earth, such that the nudge required to divert it wouldn't need to be too large. If the object were being destroyed, then the particles produced by the explosion would have time, and space, to dissipate and not rain down directly on Earth. The closer the object is to Earth, the larger the required force would be to change its trajectory adequately for it to miss the planet. If you imagine throwing a dart at a board, it's much easier to hit the target if you stand close to it. If you stand very far away from the board, then you are more likely to miss the target because a small change in the angle at which you release the dart will be amplified over the greater distance it travels. When nudging a space object, such an effect can work in our favour – the further away we perform the nudge, the smaller it needs to be to produce a large miss of our planet. However, gently nudging an object from its orbit requires at least several years' notice. In fact, a decade would be a reasonable timeframe in which to launch a mission, give it time to travel to the object and gently work some magic on arrival.

Anyway, how do we go about nudging a huge space rock onto a new orbit? Surely this can't be a simple manoeuvre when you consider that any spacecraft is going to be tiny in comparison with the object we want to move. Luckily, with enough time, even a small unmanned spacecraft could tug an asteroid or comet enough to alter its course to a more benign orbit to miss Earth. When we say 'tug', what we actually mean is that the spacecraft will travel next to the object, with the spacecraft and object mutually gravitationally attracting one another. The spacecraft can then use a small force from the use of, for example, ion thrusters, to counter the gravitational pull of the space object. The net effect is that slowly, over time, the object is moved towards the spacecraft very slightly and so is deflected from its original orbit, forever altering its course. This technique is called 'gravity tractor', and although it may be slow, one of its major advantages is that it can be used on most objects, including those of rubble-pile construction – those poorly consolidated asteroids and comets composed of a loosely bound pile of rocks (and possibly ice). A gravity tractor requires no impulsive force so it wouldn't be expected to break apart a rubble pile, meaning we would be unlikely to end up with two or more pieces of asteroid or comet heading for us, a situation we'd very much like to avoid.

There are various other, apparently gentle and slow, techniques that could act to nudge an object out of our way: these include, but are not limited to, foil wrapping or spray-painting, ion-beam shepherding, deploying a swarm of reflective mirror bees or laser ablation. The first two techniques involve contact with the object, but the others are contactless. Foil wrapping a large space object will obviously not be as easy as wrapping a potato for the oven, so it might seem a bit crazy at first, but it is a technology that scientists currently consider feasible. In practice, a special metallised plastic sheeting, not dissimilar to the blankets used by athletes at the end of endurance races or by the emergency services to keep people warm, would be used on the object to act as a solar sail. A solar sail was used on the *Messenger* spacecraft when it was travelling to Mercury, something that allowed it to harness the power of the Sun's photons, thus saving fuel. The idea is that the space object's reflective foil-wrapping shell would increase its albedo and, therefore, the degree to which it reflects the Sun's light. The result is that the effective pressure exerted by the sunlight on it changes, which acts to change the course of the object slowly over time.

Spray-painting a space object, on the other hand, may sound like a massive game of cosmic paintballing, but it could be an effective way to achieve the same result as foil wrapping. The paintballs, made either from paler titanium dioxide or darker soot, would be fired at the object to alter its albedo. In turn, the Sun's photons would bounce off its surface and change the solar radiation pressure acting on the object, nudging it in a different direction. Scientists have studied a test case of spray-painting a space object and found that an NEA called Apophis, which is 270m (900ft) in diameter, would require 5 tonnes of paint to cover it, but would also require 20 years once painted for enough solar radiation pressure to pull it from its current course. Luckily, we don't need to worry about this one as it will just be making some close passes to Earth in 2029 and again in 2036 with no cause for concern, but it is interesting to get a handle on how long it might take to divert something of this size.

The aim of mirror bees and laser ablation is the same – to heat up small regions of the space object to produce flash sublimation or vaporisation of internal gases within the object. The ejection of gas would produce a small amount of thrust and a change in its course, amplifying the Yarkovsky effect. The mirror bees would surround the object, focusing and concentrating the Sun's energy onto a small region of its surface. Laser ablation would use a focused laser beam from a spacecraft hovering nearby to heat up the object. Finally, ion-beam shepherding would direct an aligned beam of ions at the surface of the space object, acting to gently push it onto a new course. For smaller asteroids, those less than 200m (656ft) in diameter, it is thought that ion-beam shepherding might outperform a gravity tractor. These gentle techniques, despite being slow and requiring decades of advance warning of an impact, could be a sensible and safe way to realign an object to avoid it hitting Earth. If we start working on the technologies now, and start testing them in space, which is exactly what scientists are beginning to do, then we, or our descendants, will be better placed to act when we need to. But what if we spot an object that is going to collide with Earth within a few years, or earlier? Well, this is where the more destructive, faster acting, and arguably more exciting, techniques would need to step up to the mark.

Smash, grab, obliterate

It's possible that we won't know about an impending impact with Earth until very late, as sometimes the Solar System sends us a curveball to keep things interesting. This applies, in particular, to the long-period comets that can appear in the inner Solar System rather unexpectedly, on a random orbit, after departing their home in the Oort Cloud. They also travel very fast, resulting in high-impact velocities, which are, therefore, potentially more destructive. The chance of a comet of this type colliding with Earth is exceptionally low, as Earth is, after all, a very small target within the vast expanse of space. However, astronomers simply can't spot everything that's out there long enough in advance to launch a mission to divert the object.

In the event that we only had, at most, a few years' notice of an impending impact, we would need to focus on the less precise and more drastic measures of diversion or destruction. It is still the case that the earlier we act the better. If we want to blow something up, then we need to think about the shrapnel that would be produced by an explosion: how big it will be and where it will end up. Starting with the simple end of the 'drastic' scale, we could consider projecting a large kinetic impactor – like a cannon ball – towards the object to either dramatically alter its course, or completely smash it into pieces. To be blunter about it, we want to ram the thing out of the way and if that results in it breaking apart then so be it. NASA suggests that a simple kinetic impactor is the 'most mature approach' to deflecting an NEO, as long as it consists of a small, single body. However, if we were to intentionally, or accidentally, smash up the object during this process, then the outcome for Earth could be harder to predict. Depending on the composition of the space object, it might fragment into tiny dust-sized pieces that could rain down to Earth, or it might break into just a few large pieces, which, if still heading for an Earth impact, could make matters worse. In addition, a rubble pile-type space body, of which asteroid Itokawa visited by the *Hayabusa* mission is thought to be one, would more easily absorb the energy from a kinetic impact – even a large one – such that its orbit might remain unaltered. On the other hand, a similar-sized object composed of more consolidated rock would be less likely to absorb the energy of an impact and so its orbit would be expected to shift more easily

The European Union is currently working on NEOShield-2, a project to assess the technologies and preparation for NEO deflection. In test cases, it is thought that a kinetic impactor will need to be travelling at very high velocity, perhaps something like 10km/s (22,369mph), in order to deviate a NEO from its Earth-crossing course. It is important we know as much as we can about the object before we try to shift it, so we don't get any surprises such as fragmentation. So, scientists think it would be sensible to start out with a reconnaissance mission to the offending NEO to survey its surface first and to find out its exact size, shape, rotation speed and possibly even chemical composition. Once these factors are known, then the outcome of a kinetic impact would be much more predictable. Despite this, launching a reconnaissance mission and then a second mission to ram the object away is costly and time-consuming. Cost would, perhaps, be no issue since we could be saving all of our planet's inhabitants, so timeliness would be the key factor.

NASA has, of course, already trialled a kinetic impactor in space; we've seen that the *Deep Impact* mission sent a 370kg (816lb) impactor careering into the side of comet Tempel 1 in 2005. *Deep Impact* was an important test ground because it showed space agencies that they were able to accurately hit an object in space with only a small impactor, something they hadn't yet proven. Despite the impactor releasing 19 gigajoules (that's 4.8 tonnes of TNT) of energy, it didn't destroy the comet. In fact, it didn't even alter its course, but it did leave a neat pockmark in its side. The resulting impact scar is thought to be up to 250m (820ft) in diameter and 30m (100ft) deep. Tempel 1 is over 7km (4.3 miles) in its longest dimension, so it would have required a considerably larger, or more sophisticated, impactor for it to be diverted from its course, or destroyed.

At the time of writing, ESA and NASA are planning a joint mission to a moonlet of the NEA 65803 Didymos, with the aim of testing asteroid-deflection technology. Didymos is a binary asteroid measuring around 800m (2,625ft) in diameter, with its moonlet measuring just 160m (525ft). The plan is for ESA to deploy its *Asteroid Impact Mission* (*AIM*) to rendezvous with the asteroid and enter into orbit around it in 2022, with NASA following up a few months later with its *Double Asteroid Redirection Test* (*DART*) mission, which will be timed to hit the moonlet while *AIM* records the event. *DART* weighs in at around 500kg (1,100lb) and it should career into the side of the moonlet at around 6km/s (13,421mph), which is expected to alter the orbital speed of the moonlet around Didymos by about 0.5mm/s. This may sound tiny, but it will be measurable by *AIM* and ground-based telescopes. Such a mission provides the perfect test ground for assessing how such technology can work in space, even if Didymos will be the smallest asteroid ever visited.

In Chapter 9 we talked about NASA's plans to use a spacecraft to grab an asteroid and move it to a new location, either for the purposes of mining its resources or purely for scientific discovery. Although this mission and technology are, as yet, untested, if NASA can figure out how to grab a non-threatening asteroid and move it for mining purposes, then there's no reason the same approach couldn't be used on an Earth-threatening object. Only time will tell if this technique will be fruitful for NASA and the space miners out there, such that it can become a tested impact-avoidance option.

If we have very limited time, then we require an approach that can work in weeks or months. So, a more drastic method to avoid an impact with Earth, one that could work on any object regardless of its composition, might be called for. How about using a nuclear weapon? It may sound like total madness, but believe it or not, scientists are looking at the possibility of firing such a device at a space object to blow it into tiny pieces, maybe even reducing it to a cloud of gas and liquid droplets. It may sound like a great solution but there's a potentially big problem. The resultant shrapnel from such an explosion, however small, would be highly radioactive, so it's probably not something we'd want raining down on the planet. This would almost certainly be the case if we blew up the object at short notice as it was heading towards us. If the object was one that passed Earth frequently, moving ever closer to impact with each orbit, then it could be blown up in a pre-emptive strike on one of its prior close-Earth visits before the one that was predicted to cause total annihilation, nuking it as it was heading away from the Earth. In this way, any radioactive fallout from the destruction wouldn't affect life on Earth. It may sound like a relief that, whatever happens, there is a strategy to save us.

However, there are some further complications. First, according to the Outer Space Treaty, nuclear weapons are not permitted to be used in space. Maybe in an extraordinary circumstance – such as a major threat to the Earth – an exception could be made to save us from impending doom. Second, a modern-day nuclear warhead probably wouldn't survive the impact energies associated with this course of action. Instead, it would more than likely need to be detonated close to the space object for the nuclear blast energy to nudge it to a different course, otherwise known as a nuclear stand-off explosion. Alternatively, the heat from the blast could be used to disintegrate the object, which might work well if it contained a lot of volatile materials. In this scenario, we would still obviously have to think very carefully about any radioactive fallout issues.

Of course, there is an alternative that comes straight from Hollywood, and that is to place the nuclear weapons within the interior of the offending space object before they are detonated so they can fragment it from the inside out. In the movies we require the likes of Ben Affleck and Bruce Willis playing would-be space miners to risk their lives for such a cause but, in reality, robotic technology could be used. If a robot could place a nuclear warhead within an object, perhaps after excavating a hole in the side first, then it is thought it could work well to break up smaller objects that were heading our way, those less than around 400m (1,300ft) in diameter. For larger objects, a nuclear warhead would have to be placed unfeasibly deep to be successful, while risking the possibility of just breaking the object into a number of large pieces that could still be heading for Earth, multiplying the number of potentially still very large impacts expected. At least if the pieces produced during the explosion were small, around 35m (115ft) in diameter or less, then they would be expected to burn up on atmospheric entry and not cause us any harm on Earth, apart from the possible radioactivity issue.

Why should we care?

It seems there isn't just one simple approach that will work to divert or destroy any space object that could possibly be heading our way. As we've seen, space objects come in all shapes, sizes, compositions and structures, and they travel at different speeds with individual orbits. Their journeys around the Sun could mean they will intercept Earth at some point in the future. If you aren't convinced by now that studying space is important, then you probably won't have read this far anyway. But even if you are still unconvinced about the future of space exploration, then surely there is no better reason for us to study space than to avoid harm coming to the planet we call home. If we don't study space, and all the varied objects it contains, then we have little chance of identifying a threat and knowing what, if anything, can be done about it. In return, we may get to save all life on Earth, and as a freebie we get to further our knowledge of our place in the Solar System, galaxy and Universe.

Curiosity and adventure are surely innate human traits, even if they sometimes get overlooked as we go about our everyday lives of working, eating, sleeping and shopping. Humans have very short memories, we are goldfish when it comes to geological time, we forget or we simply can't fathom the amount of history experienced by our planet. We can't help it, we just don't live for long enough to really appreciate these immense timescales. However, just because in our living history we haven't witnessed a major space rock careering into the side of our planet doesn't mean that one isn't heading for us. We really shouldn't take the 'bury our head in the sand' approach, however appealing it might be for the current Earth inhabitants who are unlikely to be affected by such an event. We can't easily put a timeframe on when the next large impact will occur, as we have to talk in averages. It's useless saying that an Earth-destroying impact happens every 'X' number of years, because it then always feels like one is imminent when, compared with our relatively brief time in existence as individual humans, it isn't. The Earth exists in geological time, which is hard to imagine if we only think in blocks of one century. So, knowing we are probably safe for at least another century naturally means that we put it out of our minds and carry on with our busy lives. Plus, we have to remember that we aren't totally safe as there are always those rogue objects out there that could sneak up on us.

Of course, we can't study every single object in space – there are just too many of them – but the more we find out about the ones we can see and approach with a spacecraft, the better prepared we will be to deal with others, even those we haven't seen up close before. We can even be well-placed to deal with objects that give us very little warning of their arrival on our planet. Over the next few decades we have the opportunity to build up a detailed knowledge of space objects in our neighbourhood, and even ones that are still very far from us. We can view them remotely with ground- and space-based telescopes and we have the chance to visit some of them with spacecraft to build on these findings, providing the finer details of their composition and structure. In the future, as technological developments are made, ground-based observations might be all we need to learn everything about an object in space, even when it is a great distance away. Current telescope technology can't spot everything, and we need the ability to cross-reference telescope data with spacecraft observation and sampling to provide the ground truth. Certainly, for the time being we need spacecraft, and eventually humans, to go up there and explore the space around us in the hope that one day we can save ourselves from an impending impact. In the meantime, we can learn about where we came from, how we got here and whether we are really alone in the vastness of stars, planets and the empty vacuum of space that surrounds our tiny planet.

Epilogue

Going back to my original question: should comets and asteroids be feared or revered? Well, my answer probably won't surprise you and it is, of course, that they should be revered. It's certainly possible that human existence on our planet may one day end because of a huge Earth-shattering impact, but we must remember that life might have only begun in the first place because of very similar impacts early in Earth history. Whether the ingredients for life were delivered to Earth from space on a comet or asteroid is, as yet, unknown, but the evidence points towards this being the most likely option. However, even if comet and asteroid impacts weren't responsible for seeding the Earth with the ingredients for life, then their role in events such as the mass extinction 65 million years ago could be just one example of how they helped create an environment that allowed life like ours to develop. The end for the dinosaurs, and many other species, meant that niches were opened up for a range of mammals in different shapes and sizes to find a comfortable place on Earth, in a landscape that was suddenly free of competitors. This includes the mammals from which we are descended. What we can be certain of is that impacts from space have had an important role to play in the development of our planet, even the formation of our Moon. How, and if, they will play a role in our future is yet to be seen but we are, partly, in control of that.

We must accept that one day we will be focusing our efforts on destroying or diverting a space object intent on ending life on Earth, or we may even have the option of leaving our planet altogether to go in search of another to call home. This all sounds a bit like science fiction, but as we've seen during recent space missions, sometimes science fiction isn't too far from reality. Space agencies and commercial space companies are rapidly pushing the boundaries of what is feasible. After all, landing a functioning scientific laboratory successfully on the side of a speeding comet is something that humans can only have dreamed about just a century ago.

I truly hope that one day we won't have to leave behind our beautiful and unique planet because we've inadvertently ruined it in some way, or that we haven't figured out how to divert a space object away. The Sun has a good 5 billion years of life remaining before it eventually burns its way through its remaining stores of hydrogen. Either way, we'll be long gone by then. Nevertheless, until the Sun takes its final gasps, whatever we do the Earth will persist on its journey around the Sun with or without us and it will continue to look roughly the same, save for the continual resurfacing it experiences through geological processing. If, and when, humans one day disappear from the surface of the planet then the room they leave behind will more than likely be filled by new organisms, but not necessarily more intelligent ones!

In the book of our Solar System, we're barely halfway through, and humans have made but a fleeting appearance on just a few of its pages. Over the course of this book we've learnt that, however we arrived on this planet, we came from outer space because there is, quite simply, nowhere else we could have come from. It seems likely that the basic building blocks that make up humans, and all the other life on our planet, survived some treacherous and exciting times in history before making planet Earth their home to grow and evolve into a range of fantastic, wild and wonderful creatures. Life has made our planet the place it is, interacting with the geology and incorporating itself into the ground rock that makes up Earth. But we must remember that we are quite literally made of stardust and we will, in the far and distant future, when our Sun has taken its final breaths, return to stardust once again. At that stage, when we are ejected back into interstellar space, we might eventually be incorporated into another star. We truly are prehistoric and eternal explorers of the Universe, and the possibilities are simply endless for what we might become in the future. It might not sound very romantic, but in some ways it is beautiful. We will be around for as long as our Universe exists. Maybe, as a tiny piece of space dust we will, one day, even find our way to another welcoming planet in a far-off corner of the galaxy, to live another life.

Glossary

Albedo – The amount of light reflected by a surface, typically on a planet or moon.

Asteroid belt – A circumstellar disc of irregularly shaped space objects located between the orbits of Mars and Jupiter.

Astronomical unit (AU) – A unit equal to the distance between the Earth and the Sun; 1AU is around 149.6 million km (93 million miles).

Biosphere – The layer of planet Earth in which life exists.

Clathrate – Crystalline ice that traps non-polar molecules such as gas within a hydrogen-bonded cage-like structure.

Coma – A cloud of gas and dust surrounding a comet.

D/H ratio – The ratio between heavy hydrogen (2H) and hydrogen (1H).

Differentiated object – A space object that is large enough to have internally segregated under gravity into a layered structure composed of a core, mantle and crust (Earth being the most obvious example).

Ecliptic – The ecliptic plane includes most of the objects that orbit the Sun and is the circular path the Sun appears to follow on the celestial sphere over the course of a year.

Evolved object – Used throughout this book to refer to Solar System objects that have experienced relatively high levels of alteration near the Sun, often resulting in the transformation of their early-inherited solar nebula ingredients into more complex mineral assemblages. This category includes the rocky asteroids (inner and middle asteroid belt) and planets. Compare with 'primitive object'.

Grand-tack model – A hypothesis to account for the migration of Jupiter towards, and then away from, the Sun soon after it formed.

Heliocentric distance – Distance as measured from the centre of the Sun. Usually referred to in relative terms here (large heliocentric distances being far away from the Sun).

Heliosphere – The part of the Solar System where the solar wind has influence.

Hydrosphere – Includes the water on Earth's surface and sometimes the clouds.

IDP (Interplanetary dust particle) – Small (less than a few hundred micrometres in diameter) cosmic dust particles that are most commonly collected in the Earth's stratosphere. IDPs may originate from any Solar System body, but the vast majority are thought to be from Jupiter-family comets.

Interstellar medium (ISM) – The matter that exists in the space between stars and galaxies.

Kuiper Belt – A circumstellar disc of icy objects extending from the orbit of Neptune (around 30AU) to around 55AU.

Lagrange point – A part of space where the gravitational influence of two large bodies is equal to the centrifugal force felt by a much smaller third body.

Late-Heavy Bombardment (LHB) or Lunar Cataclysm – An intense period of Solar System formation around 4.1 to 3.8 billion years ago marked by asteroid and comet impacts, as evidenced by the cratering on the surface of the Moon.

Lithophile – 'Rock-loving'; lithophile elements combine readily with oxygen.

Meteoroid – A small body moving in space that will become a meteorite if it encounters a planetary surface.

Oort Cloud – A theoretical cloud, or shell, of icy objects (comets) in the farthest reaches of the Solar System, extending from 1,000 to 200,000AU.

Perihelion – The point at which the orbit of an object in space is closest to the Sun.

Picogram – A weight measurement unit equal to 1 trillionth of a gram.

Planetesimal – A primitive small body in the early Solar System that could, or did, become a planet.

Plate tectonics – A theory to account for the movement of the Earth's lithospheric plates, the action of which gradually destroys and re-forms the surface of the Earth and often results in earthquakes, mountain chains and volcanoes, among other features.

Presolar grains – Interstellar solid matter that originated from another star.

Primitive object – Used throughout this book to refer to Solar System objects and materials that have experienced little or no alteration near the Sun and so retain their early solar nebula chemical signatures and structures. This category includes the majority of comets and many of the asteroids present in the outer asteroid belt. Compare with 'evolved object'.

Protoplanetary disc – A rotating disc of gas and dust around a young star.

Near-Earth object (NEO) – Objects that have orbits within 1.3AU of Earth. About 1 per cent of NEOs are comets (NECs), suggesting that the population is dominated by asteroids (NEAs).

Nice model – Accounting for the dynamical evolution of the Solar System, named after the location where the scientists who initially developed the model worked (Nice, France).

Regolith – The layer of unconsolidated, loose, rocky and dusty material sitting on bedrock.

Siderophile – 'Iron-loving'; siderophile elements tend to bond with iron.

Solar nebula – The infant cloud, a swirling disc, remaining after the Sun formed from a molecular cloud.

Acknowledgments

There are so many people I'd like to thank for helping me get to this point – having completed my first popular science book. I'll start with the person who persuaded me to give geology a go in the first place, despite my scepticism that it was 'just looking at rocks'. Mr Birch, my A-level geology teacher at The College of Richard Collyer in Horsham – you may have influenced my decision by dangling the golden carrot of a field trip to Tenerife, but it worked. Little did I know that it would be the start of my career in science. Climbing through lava tubes on the side of an active volcano a few months after starting the course really got me hooked.

There are so many amazing scientists and friends whom I met throughout my degrees at Durham University and Edinburgh University who further honed my love for rocks and learning about the Earth and Solar System. Professor Colin Macpherson, my masters supervisor, you were so supportive during my degree. Your tutorials helped me with my first foray into the world of Microsoft Excel, too, and I haven't looked back. My PhD supervisor, Professor Godfrey J. Fitton, also gets a special mention, plus you taught me a thing or two about off-road driving and wine at the same time!

During the writing of *Catching Stardust,* I was thrilled to find that many of my family, friends and colleagues wanted to help me out by proofreading chapters. So many, in fact, that I couldn't take them all up on their offers, but I am still grateful they were keen to help. The following people are the 'lucky' few who read chapters and provided me with grammatical or scientific advice, which made the book better and forced me to think carefully about the message I was trying to get across. In no particular order, other than alphabetical by last name: Dr Feargus Abernethy, Gemma Allen, Dr Rosalind Armitage, Dr Jessica Barnes, Helen Cooke, Dr Richard Greenwood, Dr Lydia Hallis, Dr Claire McCleod, Dr Rhian Meara, Tess Mize, Dr Geraint Morgan,

Dr Amanda Nahm, Dr James Mortimer, Dr Ben Rozitis, Dr Graham Smith, Dr Richard Taylor, Wendy Tomlins, Clare de Villanueva and Michelle Webb.

I thank Jim Martin at Bloomsbury for his annual emails asking whether I'd consider writing a book 'about comets', to which I annually replied, 'No I have no time.' Well, I finally found the time and I'm so happy you had the faith in me to give this a go … and also for always asking how my dog was doing. To my editor, Anna MacDiarmid, who gave me useful feedback on the manuscript and was always happy to answer my many questions. Also to my amazing copy editor, Emily Kearns, who picked up so many things I missed and who generally improved the manuscript to make it what it is.

Colin Starkey gets a special mention because he remains the only person to have read my entire PhD thesis and my entire book, for which I can only apologise. *Catching Stardust* is three times longer than my thesis: you didn't even know what you were letting yourself in for. I am so happy you took the time to help me out with the first draft of every chapter, before I was happy for anyone else to see them. I can't promise this will be the last thing I ask you to proofread for me, but I blame you for being such an avid reader. Dr Michelle Starkey also gets a special mention for her biological expertise and for being my amazingly supportive sister. I'm constantly in awe of all you've achieved, and continue to achieve, and I'm astonished you had time to help me out with all that you have going on in life.

And now for the soppy bit. I thank my amazing husband. Although your job forced us to move halfway across the world, meaning that I had to leave behind the job and family I loved, it did give me the freedom to finally write *Catching Stardust*, something I'd wanted to do for years. It's been an extraordinary adventure so far, and you are the best father to our little girl that I could ever hope for. I know that with you by my side I will have many more adventures in the years ahead. I can't wait.

I thank my little girl who was literally with me – forming and growing – for the majority of the time I was writing

Catching Stardust. Even though she doesn't know she helped, the impending arrival of a child really does focus the mind. You are such a delight to be with, I love you to the Moon and back, and back again. I hope one day you'll want to move on from *The Gruffalo* to read *Catching Stardust*, but perhaps you absorbed the science in some way anyway. Also, I thank my amazing babysitter, Jamin Rexing, for caring so well for my baby when I needed to focus on finishing this book.

Finally, I thank Jacqueline Starkey for supporting me through school and life. Particularly in the years when I hated to read the books the school wanted me to read and you told me to just read whatever I liked, as long as I read something. That turned out to be *The X-Files* books instead of Jane Austen, but I like to think it didn't hold me back. Plus, in my adult years I've eventually worked my way through that school book list anyway.

Index

Sub-headings in *italics* indicate figures.